DYNAMIC METEOROLOGY: A BASIC COURSE

DYNAMIC METEOROLOGY: A BASIC COURSE

Adrian Gordon
School of Earth Sciences, Flinders University, South Australia

Warwick Grace
Bureau of Meteorology, Australia

Peter Schwerdtfeger
School of Earth Sciences, Flinders University, South Australia

Roland Byron-Scott
School of Earth Sciences, Flinders University, South Australia

A member of the Hodder Headline Group
LONDON • NEW YORK • SYDNEY • AUCKLAND

First published in Great Britain in 1998 by
Arnold, a member of the Hodder Headline Group
338 Euston Road, London NW1 3BH
http://www.arnoldpublishers.com

Copublished in the USA, Central and South America by
John Wiley & Sons, Inc.,
605 Third Avenue, New York, NY 10158-0012

British Library Cataloguing in Publication Data
A catalogue entry for this book is available from the British Library

Library of Congress Cataloging-in-Publication Data
A catalog entry for this book is available from the Library of Congress

ISBN 0 340 59503 5 (pb)
ISBN 0 340 70592 2 (hb)
ISBN 0 470 24417 8 (pb, USA only)
ISBN 0 470 24418 6 (hb, USA only)

Production Editor: James Rabson
Production Controller: Rose James
Cover design: Terry Griffiths

Composition by Alden, Oxford
Printed and bound in Great Britain by J W Arrowsmith, Bristol

CONTENTS

FOREWORD

In the first edition of this text Sir Graham Sutton, then Director General of the UK Meteorological Office and formerly Bashforth Professor of Mathematical Physics at the Royal Military College of Science, wrote in his foreword:

> A course of study in science may take one of two shapes. It may spread horizontally rather than vertically, with greater attention to the security of the foundations than to the level attained, or it may be deliberately designed to reach the heights by the quickest route possible. The tradition of scientific education in this country has been in favour of the former method, and despite the need to produce technologists quickly, I am convinced that the traditional policy is still the sounder. Experience shows that the student who has received a thorough unhurried training in the fundamentals reaches the stage of productive or original work very little, if at all, behind the person who has been persuaded to specialize at a much earlier stage, and in later life there is little doubt who is the better educated.

Although I have always agreed with these comments I did not at the time fully appreciate the real importance of their meaning.

Some 20 years later I read the cult novel *Zen and the Art of Motor-Cycle Maintenance* by Robert Pirsig. This novel strongly influenced my thinking. In the preambles to my PhD thesis I included the following extract:

> the only real learning results from hang-ups, where instead of expanding the branches you already know, you have to stop and drift laterally for a while until you come across something that allows you to expand the roots of what you already know.

Yet, I did not in any way relate, or connect, the latter quote with the ideas expressed so concisely by Sir Graham Sutton in the 1962 edition. In fact it was not until I reread the Foreword in the course of reconstructing the present text that I realized that Sir Graham was really talking about an aspect of 'lateral thinking', a concept often discussed in educational and research circles today.

The authors hope, and believe, that the tradition pioneered by Sir Graham has been carried on in this newly constructed text, comprising three fundamental courses in meteorology which have been taught to undergraduate students by the authors in dynamic, synoptic and radiation meteorology. The authors are all internationally known in their individual fields. Not only is this author diversity unusual in a book of this level, but diversity in content emphasizes the importance of the contrast in motions between the southern and northern hemispheres as determined by the differing sign of the angular velocity of rotation of the planet relative to the observer.

I hope that this text will serve its purpose well.

Adrian Gordon

ACKNOWLEDGEMENTS

The authors acknowledge the staff at the Bureau of Meteorology in Australia, especially Frank Woodcock, Noel Davidson (both for critical comments), Andrew Hollis for library assistance and Susan Wong for some word processing. Thanks also to Judy Laing for some diagram work and to Mark Bedson of Ceduna in South Australia for the cloud photos.

1

INTRODUCTION:
UNITS AND DIMENSIONS

1.1 HISTORICAL PERSPECTIVE OF METEOROLOGY

During the first part of the present century the term meteorology really meant the study of all aspects of the atmosphere. But it tended to emphasize the weather in all its variety of manifestations: the simple and the extreme, the sunny days we all enjoy, the more extreme conditions we bear and suffer and swear at. The study was particularly concerned with weather maps, or synoptic charts as they are referred to by those who prepare and analyse them, and to forecasting the weather using the weather map as a practical tool. The classic text in the early days of the century was Sir Napier Shaw's four-volume masterpiece *Manual of Meteorology*, followed in the 1930s by Sir David Brunt's *Physical and Dynamic Meteorology*. Even in those comparatively early days the role of mathematics was indispensable. Yet, it often seemed most difficult to follow the equations. Meteorology was only taught at a few universities – Imperial College in London, MIT (Massachusetts Institute of Technology) in the USA in the English-speaking world, but it was also advancing rapidly in Scandinavia and other European countries. It was mainly taught as a postgraduate specialist course leading to Master and PhD degrees.

In the middle 1930s, in a time of recession and depression, meteorology received a big boost. This was due to the demands of the rapidly expanding airline industry, particularly in the USA. Jobs were offered to qualified meteorologists to prepare flight weather forecasts for airline operations. A few years later the outbreak of World War II intensified the demand for weather forecasters. Training courses were established and filled by those who only had a rudimentary background of mathematics and physical science. During the latter half of this century the term meteorology tended to be displaced by the more erudite term 'atmospheric physics' or 'atmospheric science'. This transformation was in part due to a certain elitism that meteorology was too much concerned with weather maps and weather forecasting which, before the days of numerical prediction models, were thought by mathematicians and physicists to be a kind of technical trade beneath their professional abilities. But in this work we will not show disrespect to the original term, which is still useful when we wish to subdivide the subject into more specialized fields of interest.

Thus a main field of interest here is called dynamic meteorology. The word dynamic is used a great deal when it is desired to convey that something has energy and movement. It is defined in the dictionary as motive force. Thus, dynamic meteorology applies our knowledge of mathematics and of physical processes to explain and describe the motions and energy transformations which occur in the atmosphere to produce our weather, and eventually, our climate. In turn it is convenient to subdivide dynamic meteorology into headings which concern the thermodynamics and the dynamics of the atmosphere. The former concerns the effect of the influence of heat in its various forms on the vertical and horizontal structure of the atmosphere while the latter concerns and describes the resulting motions.

But we will also touch on synoptic meteorology. This heading covers the scientific techniques used in forecasting the weather by means of the analysis of synoptic charts of the surface and upper level patterns of barometric pressure, temperature and humidity. A further subdivision is that of physical meteorology which deals among other things with the heat budget, balancing the short-wave radiation directly received from the sun with the long-wave radiation emitted and received by the earth's surface, by clouds and by the atmosphere itself. This balance determines the mean temperature of our planet.

Further branches of meteorology include mesometeorology, which covers the study of meteorological processes and motions on a scale of the size of local topographical or surface features or of thunderstorm or squall line size, and micrometeorology, which covers processes and motions on the scale of a few metres or even centimetres above the land or sea surface.

Finally, the heading which has a long classical history, but which has recently undergone a rebirth to become the most important of all subheadings, because it includes all, is climate. From its origins of presenting maps of mean temperature and other observations for different parts of the world it now embodies all of the physics and mathematics of the other branches in the large numerical climate models which not only reproduce the actual climate of the planet, but attempt to predict future climates within the broad concept of 'climate change'.

Once again, meteorology has a promising future. Now it is in the forefront of discussion because it concerns the environment. The weather is part of our environment. Climate change is in the minds of everyone because it directly affects everyone. Hence the need for the scientific issues to be properly understood. These issues all stem from the basic mathematical and physical laws which govern our universe. These basic laws will be derived in an easy, step-by-step manner and presented in the form of simplified mathematical equations.

1.2 DIMENSIONS

In science generally, and in mathematics and physics particularly, it is most important at all stages to keep track of the dimensions of the quantities being manipulated. There are important practical reasons for doing this apart from theoretical rigidity. For example, the reader may be finding the answer to a problem which involves the derivation of a complicated set of equations. After pages of work the state of the equations may seem to be leading to an impossible solution. Rather than go back to the beginning and repeat all the operations a simple check may be made of the dimensional value of each individual term and expression on both sides of the equal sign in the equation. If any one term has a dimension which differs from the remaining terms there is an error somewhere in that term. The error can then be found more easily by tracking the earlier development of that particular term.

In meteorology there are four dimensions of which the first three are fundamental. These are usually denoted as

M   mass
L   length
T   time

A fourth dimension appears in some physical quantities. It is denoted as

K   temperature

The use of the capital letter K arises from the fundamental unit of temperature, the kelvin. The kelvin scale of temperature starts at 0 K or absolute zero, at which value there would in theory be a total absence of any heat energy in the domain in which the temperature is being assessed.

All physical entities must have a dimension associated with them. Some quantities, such as the ratio between two quantities having the same dimensions, are non-dimensional or dimensionless. These are just pure numbers.

Table 1.1 lists some of the more common quantities which occur in traditional physics and therefore in meteorology, which is a branch of physics. The first column gives the quantity, the second column lists the dimensions of the quantity and the third column the method of expressing the numerical value of the quantity, which we will discuss in the following section.

In order to understand the dimensions of the various quantities listed it is necessary to appreciate the simple physical relationships such as

Force = mass × acceleration $F = Ma$
Pressure = force per unit area $P = FA^{-1}$
Work = force × distance $W = FL$
Work = energy = heat $W = E = H$
Power = work ÷ time $P = WT^{-1}$

and other basic statements learned in elementary school physics.

We will now discuss the entries in the third column of Table 1.1.

TABLE 1.1 Dimensions and units of quantities used in meteorology

Quantity	Dimension	Units (SI)
Area	L^2	m^2
Volume	L^3	m^3
Density	M/L^3	$kg\,m^{-3}$
Specific volume	L/M^3	$m^3\,kg^{-1}$
Velocity	L/T	$m\,s^{-1}$
Acceleration	L/T^2	$m\,s^{-2}$
Force	ML/T^2	N
Pressure	M/LT^2	Pa
Work	ML^2/T^2	J
Energy	ML^2/T^2	J
Angular velocity	$1/T$	$rad\,s^{-1}$
Momentum	ML/T	$kg\,m\,s^{-1}$
Divergence	$1/T$	s^{-1}
Vorticity	$1/T$	s^{-1}
Power	ML^2/T^3	$J\,s^{-1}$
Frequency	$1/T$	$cycles\,s^{-1}$
Wavelength	L	m
Lapse rate	K/L	$K\,m^{-1}$
Specific heat	L^2/KT^2	$J\,kg^{-1}\,K^{-1}$
Latent heat	L^2/T^2	$J\,kg^{-1}$

1.3 UNITS

In the past units have been defined in the c.g.s. or centimetre, gram, second system. In this system the unit of force is the dyne, the force required to give a mass of 1 gram an acceleration of $1\,cm\,s^{-1}\,s^{-1}$. The most commonly used unit of pressure in this system is the millibar (mb) which is $1000\,dynes\,cm^{-2}$. The unit of work is the erg. In recent years the c.g.s. system has been replaced by the m.k.s. or SI system (International System), in which the unit of force is the newton (N), the force required to give a mass of 1 kg an acceleration of $1\,m\,s^{-1}\,s^{-1}$. The unit of pressure is the pascal (Pa) which is $1\,N\,m^{-2}$. The unit of work is the joule. We will, however, retain the unit of millibar in our discussions of synoptic charts and upper air diagrams as this unit tends to be used more widely than kilopascals or hectopascals (hPa). One hPa is numerically equivalent to 1 millibar so nothing is lost by retaining the name of millibar. One watt is $1\,J\,s^{-1}$. We are accustomed to think of a watt as a unit of electrical energy. Electric light bulbs are labelled in watts, which measures their brightness, and is a guide to the amount of electricity they use and so to the amount on the electricity bill we have to pay. In meteorology and particularly in climate change sensitivity studies we will find that the energy received from the sun is also measured in watts. This is an example of how different branches of science interrelate with one another.

It is essential that all numerical quantities be labelled with their correct units. An answer to a worked problem is not right unless it is expressed in the proper units of measurement. The third column in Table 1.1 shows the manner in which numbers should be identified by their units. If the units are correct so are the dimensions.

The c.g.s. or, preferably, SI units must be used for all mathematical relationships. Occasionally, for practical observational purposes, it may be more convenient to use non-standard units such as knots for wind speed and degrees Celsius for temperature. Degrees Fahrenheit and inches of rain are still widely quoted in some countries.

Worked Example

You vaguely remember that the pressure (p in Pa) of unit mass (1 kg) of a perfect gas depends upon its specific gas constant (R in J kg^{-1} K^{-1}), density (ρ in kg m^{-3}) and absolute temperature (T in K), but you have forgotten the exact form of the perfect gas equation and now wish to reconstruct it!

Solution:

If [ ] denotes the dimensions of a quantity, then we have that

$$[p] = [\text{force per unit area}] = MLT^{-2}L^{-2} = ML^{-1}T^{-2}$$
$$[R] = ML^2T^{-2}M^{-1}K^{-1} = L^2T^{-2}K^{-1}$$
$$[\rho] = ML^{-3} \quad \text{and} \quad [T] = K$$

Now suppose that the perfect gas equation has the form $p = R^q \rho^r T^s$, where q, r and s are to be found. Since the dimensions of each side of a physical equation must be identical, it follows that

$$[p] = [R]^q [\rho]^r [T]^s$$

and hence that

$$ML^{-1}T^{-2} = M^r L^{2q-3r} T^{-2q} K^{s-q}$$

Equating the indices of M yields $r = 1$,
of L yields $2q - 3r = -1$, from which $q = 1$,
of T yields $-2q = -2$, from which $q = 1$ (as above),
of K yields $s - q = 0$, from which $s = 1$.

It therefore follows that the perfect gas equation must have the form $p = \rho RT$.

Note that this is not a physical proof of the perfect gas equation, for the latter must be derived from Boyle's and Charles' laws.

1.4 PROBLEMS

1. What are the dimensions and units (SI) of

    (a) wind speed
    (b) the velocity of light

    (c) the rotation of the earth on its axis
    (d) a kilowatt hour of electricity
    (e) the logarithm of the absolute temperature (K)
    (f) a number
    (g) salinity in the ocean
    (h) the concentration of carbon dioxide in the atmosphere
    (i) space
    (j) the density of a 'black hole'?

    Note: In some of the above it may be convenient to express the answer also in units of larger scale.

2.   In micrometeorology (and aerodynamics), it is assumed that the friction velocity u_* depends upon the stress τ (in $N\,m^{-2}$) and the density ρ (in $kg\,m^{-3}$) in the fluid. Use variable dimensions to find a suitable relationship between u_*, τ and ρ. [Hint: Suppose that $u_* = \tau^r \rho^s$.]

3.   Show that

    (a) 1 joule is equivalent to 10 million (10^7) ergs
    (b) 1 newton is equivalent to 10^5 dynes
    (c) 1 pascal is equivalent to 10 dynes cm^{-2}
    (d) 100 pascals (1 hPa) $= 1$ millibar.

If you can easily master the transformations in problem 3 we may proceed to the next chapter.

2

THE THERMODYNAMICS OF DRY CLEAN AIR

The atmosphere is composed of a mixture of gases which in more popular language is called air. It is made up of the entire body of gaseous substances which cover our planet earth. The lower boundary is marked by land or liquid water substances, the continents, oceans, lakes and rivers, and ice and snow surfaces. The upper boundary extends into the fringes of outer space, but at great heights the mass of the atmosphere is too small to be of consequence to our weather. The region where weather occurs is the lower part of the atmosphere distinguished by the important property that on a broad scale temperature decreases with height. This region is called the troposphere and it is the region which mainly concerns the meteorologist. Above the troposphere is the stratosphere which possesses the property that the temperature no longer decreases with height, but remains the same (isothermal) or increases a little with height. The discontinuity, or narrow zone, which divides the troposphere from the stratosphere is called the tropopause. Upper air ascents which record the temperature at different heights show a well-marked discontinuity at the tropopause.

In the troposphere the mixture of clean air consists of approximately 78% nitrogen and 21% oxygen by volume. The remaining 1% is made up of argon, carbon dioxide, and other gases. The concentration of carbon dioxide (CO_2) has been increasing owing to human activities. When the first edition of this book was published in 1962 the concentration of CO_2 was about 315 parts per million by volume. It is now about 355 ppmv. CO_2 in these quantities is not in itself a harmful gas, although an increase in concentration will probably eventually cause a global warming. The magnitude of any such warming is at present a question of some debate because of the compensating factors of clouds, and the effect of aerosols (small particles) and of volcanic eruptions. Much more research is needed to give a reliable answer to the question of predicting a reliable increase in mean temperature for the planet for the decades ahead of us.

However, we cannot be so easy on the release of other gases into the atmosphere through human activities. Sulphur dioxide (SO_2) reacts with water (H_2O) to form sulphuric acid (H_2SO_4), which falls out as acid rain. Carbon monoxide, exuded from automobile exhausts, is lethal in quantity. Nitrous oxide, methane, and the chlorofluorocarbons (CFCs) are other obnoxious gases which are poured into the atmosphere. What appears as pollution in the atmosphere is composed of particulate matter, small particles of soot (solid carbon), sulphur compounds and other chemicals. In the last century, and up to the middle of the present century, London was renowned for its 'pea soup' fogs, caused by the burning of coal. These fogs smelled acrid and were harmful to the human respiratory system. In December 1952 more than a thousand deaths were attributed to a prolonged spell of such fog during a quiet period of anticyclonic weather. In consequence laws were passed to prevent the burning of fuels which caused such disastrous effects on the atmosphere and London became free from the kind of fogs described in the opening page of Charles Dickens' *Bleak House*.

However, the concentrations of effluent harmful gases are not large enough to affect the broad-scale thermodynamics or dynamics of the atmosphere. There is one other constituent of the atmosphere which must be mentioned at this stage, and that is water vapour. Water vapour is also a strong 'greenhouse gas'. Its globally averaged concentration throughout the whole atmosphere is approximately 2.5 grams kg^{-1}, that is about 0.25%, which is about one four-hundredth of the total mass of air. It can be seen from the previously stated concentration of CO_2 that there is on average seven times more water vapour than CO_2. In the surface layer in the tropics there is 45 times more water vapour than CO_2. We will discuss water vapour in more detail in the next chapter. However, in deriving the thermodynamical equations we will consider clean dry air in the sense that it does not contain any water substance, solid or liquid particles, but is composed entirely of the elements mentioned above and listed in Table 2.1.

2.2 THE SCIENTIFIC METHOD

The method in which scientific knowledge is gathered has two components. The first is observational. Observations of what happens in the natural world are

Table 2.1 Molecular weights and specific gas constants of components of dry air

Gas	Mol. weight m_k	Gas constant R_k	Part by mass M_k	$M_k R_k$
Nitrogen	28.016	296.74	0.7552	224.10
Oxygen	32.000	259.80	0.2315	60.14
Argon	39.944	208.13	0.0128	2.66
Carbon dioxide	44.010	188.90	0.0005	0.09
Dry air			1.0000	286.99 = R_d

made, collected, and put in some kind of methodical order. Most observations are made by some mechanical or electronic instrument or piece of equipment. They may be made from observing the behaviour of the natural world or they may be obtained from the results of artificially contrived experiments in laboratories. The second component is analysis of the collected observations and the search for a relation between individual observations within a space–time framework; that is, within the dimensions defined in the previous chapter. Such relationships may be called laws. Thus, it may be observed that objects falling within a vacuum under the influence of gravity fall with a constant acceleration which may be measured as g, about $9.8 \, \mathrm{m \, s^{-2}}$. From this result a law may be postulated and this law may be expressed by a number of relationships, that is by mathematical equations, such as

$$\text{velocity} = \text{acceleration} \times \text{time} \quad \text{or} \quad \frac{\mathrm{d}s}{\mathrm{d}t} = gt$$

$$\text{distance fallen} = s = \tfrac{1}{2}gt^2$$

and so forth.

As we progress with our study of meteorology we shall find that there are a number of laws which have been established as a consequence of observation and/or controlled experiment. These laws, expressed as equations, govern all the complex processes which occur in the atmosphere. They determine the continuous evolution of the weather, from the hour-by-hour development and decay of cumulus clouds, to the changing patterns shown on daily weather maps and satellite cloud images, to the seasonal changes of summer and winter and the long-term evolution of the climate.

It is important that we understand the laws which are at work. This understanding can best be attained by following the derivation of the mathematical equations, or language, by which the laws are described. It is the opinion of the authors that this approach is more rewarding than accepting an equation on trust, just because it appears in print, or avoiding equations altogether and simply accepting descriptive expositions of the dynamics. The latter approach is not an acceptable alternative to a prospective career in atmospheric science.

2.3 THE EQUATION OF STATE OF A PERFECT GAS

According to kinetic theory, fluids consist of millions of molecules moving randomly and colliding often with one another and sometimes with the molecules of their boundary. In the denser fluids, which are known as liquids, the molecules take up a significant proportion of the space occupied by the fluid and they are sufficiently close together (on the average) for the forces between them to be easily called into play. At a certain critical distance, the intermolecular forces between two molecules are zero but, at greater or lesser distances, very large attractive or repulsive forces occur between the molecules. Thus, if any attempt is made to compress or decompress a liquid (i.e. to force the molecules closer or further apart, on the average), enormous intermolecular forces of repulsion or attraction tend to resist it and the liquid is said to be almost incompressible.

In the less dense fluids, which are known as gases or vapours, the molecules tend to be about 10 times further apart than in a liquid. They therefore take up very little of the space occupied by the fluid and they are so far from one another (on the average) that only very weak forces of attraction occur between molecules over most of their random motions. As a consequence of this, gases and vapours are easily compressed. When a gas or vapour is so rarefied that the proportion of space occupied by the molecules and the attractive forces between the latter are negligible, we say that we are dealing with a perfect gas. Of course, no real gas can be exactly perfect but, under natural conditions, the mixture of gases which we refer to as air is sufficiently close to perfect for most meteorological purposes.

Since even the behaviour of something as idealized as a perfect gas can only be described by mathematics, we must now derive the equation of state for a perfect gas, which is the first of a series of fundamental equations with which we must become familiar. It involves three variables which we will call p, T, α. The notation T now signifies absolute temperature in K, and must not be confused with the dimension of time, also denoted by T. p is pressure in Pa, and α is the specific volume. Specific volume is volume per unit mass. The symbol for density is ρ. Thus,

$$\alpha = 1/\rho$$

The derivation of the equation of state for a perfect gas depends on combining the results of two experimental laws. In science 'laws' or relationships between variables must first depend on experimental observations. The first experimental law we use is called Boyle's law. It states that if the temperature (in K) is held constant the volume is inversely proportional to the pressure. It merely says that if the temperature is held constant and a gas is compressed by increasing the pressure then the relation $p\alpha = $ constant is true for all stages of the process. The second experimental relation we use is Charles' law, which states that if the pressure is held constant the specific volume is directly proportional to the temperature. This law merely says that if the pressure is held constant and a gas is heated, the gas will expand and the specific volume increase in proportion to the increase in T.

We now combine these two laws in the following manner:

$$p\alpha = \text{constant} \tag{2.1}$$

$$p_1\alpha_1 = p_2\alpha_2 = p_2\alpha_{(p_2,T_1)} \tag{2.2}$$

The bracket under α_2 denotes that α_2 is at pressure p_2 and temperature T_1. Now, from equation (2.2) we know that

$$\alpha_{(p_2,T_1)} = \frac{p_1\alpha_{(p_1,T_1)}}{p_2} \tag{2.3}$$

The above expression gives the specific volume at the original temperature T_1. The process was therefore carried out while maintaining a constant temperature.

We now introduce Charles' law which states that

$$\frac{T}{\alpha} = \text{constant}$$

$$\frac{T_1}{T_2} = \frac{\alpha_{(p_2,T_1)}}{\alpha_{(p_2,T_2)}} \qquad (2.4)$$

Then

$$\alpha_{(p_2,T_1)} = \frac{T_1\alpha_{(p_2,T_2)}}{T_2} \qquad (2.5)$$

The above relation expresses the specific volume at constant pressure p_2. That is, the process took place while keeping the pressure constant at p_2. Equating the two relations (2.3) and (2.5) we have

$$\frac{p_1\alpha_{(p_1,T_1)}}{p_2} = \frac{T_1\alpha_{(p_2,T_2)}}{T_2}$$

Therefore

$$\frac{p_1\alpha_1}{T_1} = \frac{p_2\alpha_2}{T_2}$$

or

$$\frac{p\alpha}{T} = \text{constant} \qquad (2.6)$$

The constant may be determined experimentally by measuring the volume occupied by unit mass of the gas at some selected pressure and temperature. It is called the specific gas constant and is found to be $287\,\text{J}\,\text{kg}^{-1}\,\text{K}^{-1}$ for dry air. We may now write the important relation

$$p\alpha = RT \qquad (2.7)$$

Equation (2.7) is known as the equation of state of a perfect gas, referred to as the equation of state. However, in using Charles' law to derive the equation of state it is important to recognize that the law only holds in the form of (2.7) for those ranges of temperature and pressure for which the substance is in a gaseous state. If, for example, the temperature is decreased beyond a certain limit the gaseous state of a gas will be transformed to a liquid or solid state, for which the equation of state is not valid. The best example of this is water, which freezes to a solid state (ice) at atmospheric pressure (about 1013 hPa or mb) at approximately 0°C, and boils at atmospheric pressure at about 100°C. Students will be familiar with school experiments with liquid air, and everyone is familiar with 'dry ice' or solid carbon dioxide.

Equation (2.7) is the general form of the equation of state. For the case of dry air it becomes

$$p\alpha = R_d T \qquad (2.8)$$

The equation of state is one of the basic equations used throughout meteorology. The derivation we have worked through here is rigorous, but it may seem a little complicated for the first mathematical relation developed. Succeeding derivations will in many cases be simpler than this first one.

Worked Example
What is the density of a sample of dry air at the 500 hPa level if the temperature is $-20°C$? Note that we will denote the gas constant for dry air as R in the examples.

Solution:
The equation of state is

$$p\alpha = RT$$

$$\rho = \frac{p}{RT}$$

$$\rho = \frac{500 \times 100}{287 \times 253} = 0.6886 \,\text{kg m}^{-3}$$

Remember that 1 mb (hPa) is 100 Pa, where 1 Pa is the unit of pressure in the SI system. Pressure must always be expressed in equations in this way.

2.4 THE UNIVERSAL GAS CONSTANT

We have so far used two experimental laws, Boyle's and Charles', to derive a mathematical relation, the equation of state. Another law which has been established by experiment and observation was formulated by Avogadro. He found that the molar volume of a gas at the same pressure and temperature was the same for all permanent gases. The molar volume is the volume occupied by a mass of gas equal to unit mass multiplied by the molecular weight of the gas. Thus a 1 gram molecule is m grams where m is the molecular weight. The molar volume is dependent on the pressure and temperature and so, multiplying both sides of (2.8) by m,

$$pm\alpha = mR_{\text{d}}T = R^*T \tag{2.9}$$

But $m\alpha = V$, the molar volume (α is the specific volume, i.e. volume per unit mass) which is the same for all gases. We will let $mR = R^*$ where R^* is called the universal gas constant and is $8313.6 \,\text{J kg}^{-1} \,\text{mol K}^{-1}$. Substituting for R in (2.7) we obtain the form

$$p\alpha = \frac{R^*T}{m} \tag{2.10}$$

2.5 MIXTURE OF GASES

A fourth law based on experiment is known as Dalton's law of partial pressures. It states that in a mixture of perfect gases each gas completely occupies the volume; each gas obeys its own equation of state; and the sum of the partial pressures of each individual gas equals the total pressure of the mixture.

If there is a mixture of different gases such that V cubic metres contains M_1 kg of one gas, M_2 kg of a second gas and finally M_s kg of another gas which have molecular weights $m_1, m_2, \ldots, m_s$ and specific gas constants $R_1, R_2, \ldots, R_s$, where each $R_k = R^*/m_k$, then Dalton's law states that each constituent gas will obey its equation of state as though the other constituents were not present. If the partial pressures are $p_1, p_2, \ldots, p_s$

$$p_k V = M_k R_k T \quad k = 1, 2, \ldots, s \tag{2.11}$$

and

$$\sum p_k = p$$

Summing the above equation, (2.11),

$$pV = \sum_{k=1}^{k=s} M_k R_k T$$

If M denotes the total mass of the mixture and R is chosen such that

$$MR = \sum_{k=1}^{k=s} M_k R_k \tag{2.12}$$

then

$$pV = MRT$$

$$p\alpha = RT$$

which is the same as (2.7).

Thus, if R is defined as above, a mixture of perfect gases will have the same equation of state as one perfect gas by itself. Formula (2.12) states that R is simply a weighted average of all the R_k, each R_k being weighted according to the mass of gas M_k present in the mixture.

2.6 MOLECULAR WEIGHT OF DRY AIR

A value for the specific gas constant for dry air may be obtained by considering the molecular weights and specific gas constants of the constituent gases shown in Table 2.1. The gas constants for the different gases in the atmosphere are found from the relation stated in the previous section, $R_k = R^*/m_k$.

It is seen that the sum of the $M_k R_k$ values is 287 approximately. We may then define the molecular weight of dry air by use of the formula $mR = R^*$ which was obtained from (2.9):

$$m_d = \frac{R^*}{R_d} = \frac{8313.6}{287.0} = 28.97$$

With this definition equation (2.8) may be used for dry air just as for any hypothetical perfect gas of molecular weight m_d. In particular, we may consider a mixture of dry air and water vapour as a perfect gas, using the value of R appropriate to the mixture.

2.7 WORK

When a material particle under the action of a force F moves through the distance ds in the direction of the force, the element of work dW done by the

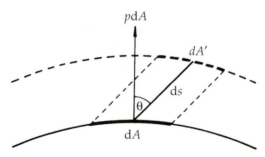

FIGURE 2.1 Definition of work.

force is $F\,ds$. When the direction of movement makes an angle θ with the force, only the displacement $ds\cos\theta$ in the direction of the force contributes to the work, and the element of work

$$dW = F\,ds\cos\theta \tag{2.13}$$

We may consider the case of the amount of work done by a gas which expands against its environment (Fig. 2.1). In considering an element of compressible fluid such as a gas the force $F = p\,dA$ where dA is an element of area upon which the force is acting. The pressure p is the force per unit area exerted by the fluid element on its boundaries.

Then

$$dW = p\,dA\,ds\cos\theta = p\,dV \tag{2.14}$$

dV is the element of volume swept through as the element of boundary area dA moves through ds to dA'. dW is thus the element of work done by the parcel of gas as it expands its boundary from dA to dA'.

The sign convention is that if work is done on the environment by the parcel dW is positive. If work is done by the environment on the parcel dW is negative. The environment is the mass of fluid surrounding the parcel under consideration.

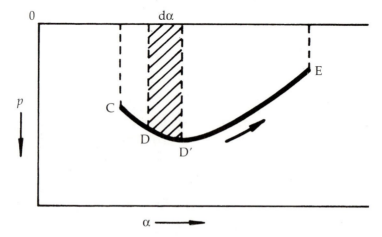

FIGURE 2.2 Path of an elementary process.

Now the state of a gas may be represented by means of a diagram with coordinates α, p in which pressure decreases upwards along a linear scale. We may follow on this diagram the elementary processes by which the gas changes its state as defined by successive pairs of values α, p. Each point on the diagram represents a unique state.

Any change of state from α, p to $\alpha + d\alpha$, $p + dp$ is called an elementary physical process. A finite process is composed of a succession of elementary ones and can be represented on the diagram by a continuous line, the path of the process. Suppose that a perfect gas changes its state from that represented by point C to that represented by point E in Fig. 2.2 by way of the various states represented by all the points of which the curve CDD'E is composed.

In Fig. 2.2 the element of work $dw = p\,d\alpha$ from (2.14), for unit mass of the gas. This equals the area of the shaded strip. The width of the element $d\alpha$ is, of course, very small compared with the length of the strip.

Then $w = \int_C^E p\,d\alpha$ for the whole process represented by the path CDD'E; it is equal to the area bounded by the curve CDD'E, the specific volume isopleths through C and E, and the upper boundary of the diagram, where p is considered to be zero.

Figure 2.3 shows a cyclic process, so called because the gas eventually returns to its initial state by way of a cycle of different states. Negative work is done on the return path EGC since the element is being compressed and work is being done on it. In this case the area bounded by the curve EGC and the upper boundary, representing the negative work, must be subtracted from the total area above CHE, representing the positive work, to obtain the area equivalent to the work done by the element. This area is that enclosed by the cyclic curve CHEGC. Thus

$$W = \oint p\,d\alpha = A \tag{2.15}$$

where the integration is performed around the closed path, and A is the area enclosed.

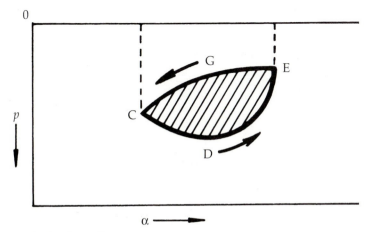

FIGURE 2.3 Path of a cyclic process.

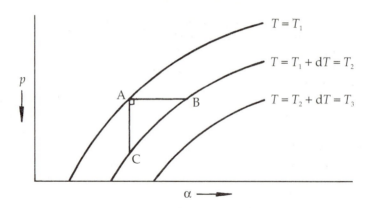

FIGURE 2.4 Curves of state.

From the equation of state $p\alpha = RT$ there must be a different curve of state for each temperature as shown by the appropriate isotherm.

Suppose it is desired to heat the system in Fig. 2.4 from temperature T_1 to temperature T_2. This can be done in an infinite number of ways. The process AB is isobaric (constant p) while the process AC is isosteric (constant α). It is noted that no work is done in the isosteric process. The process BC is isothermal (constant T).

Worked Example
How much work is needed to lift 1 kg of dry air from 1000 mb (hPa) to 900 mb (hPa) if the temperature throughout the 1000–900 mb layer is 10°C?

Solution:
Work is defined as

$$w = \oint p\,d\alpha$$

We can find this as follows.
Differentiate the equation of state

$$p\,d\alpha + \alpha\,dp = R\,dT = 0$$

since the temperature remains constant. Integrate between the required limits

$$\int_{1000}^{900} p\,d\alpha = \int_{1000}^{900} -RT\,\frac{dp}{p} = -RT\left[\log p\right]_{1000}^{900}$$

$$W = 287 \times 273 \times \log(1.111) = 8479\,\text{J}$$

2.8 HEAT

Heat must not be confused with temperature. Temperature measures how hot a substance is on the kelvin scale. The temperature may range from absolute zero to millions of degrees in the centres of hot stars. Now it is observed that when two substances with different temperatures are brought into contact with one another the warmer substance gets cooler and the cooler substance gets warmer. Heat is a form of energy and may be converted to work. It is therefore expressed in joules. In the old c.g.s. system the unit of heat is expressed as a calorie, which is defined as the amount of heat required to heat 1 gram of water from 14.5 to 15.5°C. One calorie equals 4.185 J. In the m.k.s. system it therefore takes 4185 J to heat a kilogram of water 1°C. If dT is the change in temperature of a mass of a substance and dQ the amount of heat required to effect the change then dQ/dT is defined as the heat capacity of the substance. If dT is the change in temperature of unit mass of a substance, say 1 kg, then dq is the amount of heat imparted to unit mass of the substance to effect the change. dq/dT is defined as the specific heat of the substance. Its units are $J\,kg^{-1}\,K^{-1}$.

2.9 THE FIRST LAW OF THERMODYNAMICS

The first law of thermodynamics states that mechanical energy or work and heat are equivalent to each other and so may be converted from one form to the other. In a classical experiment in 1849 Joule produced heat by churning water. He found the relation that 1 calorie = 4.185 J as already stated. As for Boyle's law, Charles' law, Avogadro's law and Dalton's law we establish a new law as a result of a physical experiment in the laboratory. It is expressed

$$dQ = dU + dW \tag{2.16}$$

This is a fundamental relationship in dynamic meteorology. We have already defined dQ and dW. The quantity dU remains to be defined.

U is the internal energy of the system and dU represents a change in that internal energy. U is a measure of the random molecular excitation and can be shown to be dependent only on temperature. The relation (2.16) states that an amount of heat added to or subtracted from a substance is used partly in changing its internal energy and partly in doing work against external pressure forces.

If (2.16) is divided by the mass of the system we obtain

$$dq = du + dw \tag{2.17}$$

and this relation refers to unit mass of the system, where lower case letters have replaced capitals in our notation.

One may also write (2.17) in the form

$$dq = du + p\,d\alpha \tag{2.18}$$

2.10 SPECIFIC HEATS OF GASES

The amount of heat required to change the temperature of a gas by any amount depends on the conditions under which the change takes place. If the gas expands during the warming some of the heat supplied will be used to do work on the environment and so more heat will be required to warm the gas by the specified amount than if the gas were at constant volume. Each process curve between T_1 and T_2 in Fig. 2.4 represents a different specific heat. Of the infinite number which are possible those represented by the processes AB and AC are of special interest. The latter specific heat is at constant volume and the former at constant pressure. Thus:

1. specific heat at constant volume

$$c_v = \left(\frac{dq}{dT}\right)_v \tag{2.19}$$

2. specific heat at constant pressure

$$c_p = \left(\frac{dq}{dT}\right)_p \tag{2.20}$$

Now if the process is isosteric, that is if it takes place at constant volume, (2.18) becomes

$$dq = du \tag{2.21}$$

$$\left(\frac{dq}{dT}\right)_v = \frac{du}{dT} = c_v \tag{2.22}$$

Thus,

$$du = c_v \, dT \tag{2.23}$$

and

$$dq = c_v \, dT$$

It is seen therefore that for an isosteric process all heat added to or taken away from a system goes to increase or decrease the temperature. If the process is not isosteric (2.18) becomes

$$dq = c_v \, dT + p \, d\alpha \tag{2.24}$$

This is the energy equation for a perfect gas.

If we differentiate (2.7), for a perfect gas

$$p \, d\alpha + \alpha \, dp = R \, dT \tag{2.25}$$

and substituting in (2.24)

$$dq = c_v \, dT + R \, dT - \alpha \, dp \tag{2.26}$$

Now consider an isobaric process; $dp = 0$ so that (2.26) reduces to

$$dq = (c_v + R) \, dT$$

or

$$\left(\frac{dq}{dT}\right)_p = c_v + R = c_p \tag{2.27}$$

which is known as Mayer's Formula. c_p must be greater than c_v since any heat imparted to a gas at constant pressure must be used to expand the gas against the environment and thus do work as well as increase its temperature. The values of the specific heats for dry air are found to be

$$c_v = 717 \, \text{J} \, \text{kg}^{-1} \, \text{K}^{-1}$$

$$c_p = 1004 \, \text{J} \, \text{kg}^{-1} \, \text{K}^{-1}$$

Worked Example
The air in a room of $100\,m^2$ and ceiling $5\,m$ from the floor is at a pressure of $1000\,mb$ (hPa) and temperature $0°C$. How much energy would be required to heat all the air in the room to $20°C$?

Solution:
The volume of the room is $500\,m^3$.
The density $= p/RT = 1.276\,kg\,m^{-3}$
The total mass is $500 \times 1.276 = 638.15\,kg$.

$$\text{Heat} = 638.15 \times 20 \times 717 = 9.15 \times 10^6\,J$$

Equation (2.24) may be written in the form

$$dq = (c_p - R)\,dT + p\,d\alpha$$

or from (2.25)

$$dq = (c_p - R)\,dT + R\,dT - \alpha\,dp$$

from which

$$dq = c_p\,dT - \alpha\,dp \qquad (2.28)$$

If we divide (2.24) and (2.28) by T

$$\frac{dq}{T} = \frac{c_v\,dT}{T} + \frac{p\,d\alpha}{T} = \frac{c_p\,dT}{T} - \frac{\alpha\,dp}{T} \qquad (2.29)$$

whence it follows that from the equation of state (2.7)

$$\frac{dq}{T} = \frac{c_v\,dT}{T} + \frac{R\,d\alpha}{\alpha} = \frac{c_p\,dT}{T} - \frac{R\,dp}{p} \qquad (2.30)$$

and

$$\frac{dq}{T} = d(\ln T^{c_v}\alpha^R) = d(\ln T^{c_p}p^{-R}) \qquad (2.31)$$

To obtain an expression for dq/T which does not involve temperature the equation of state may be differentiated logarithmically, giving

$$\frac{dp}{p} + \frac{d\alpha}{\alpha} = \frac{dT}{T} \qquad (2.32)$$

From (2.32) we can substitute for dT/T in (2.30). Then

$$\frac{dq}{T} = c_v\left(\frac{dp}{p} + \frac{d\alpha}{\alpha}\right) + R\frac{d\alpha}{\alpha} = (c_v + R)\frac{d\alpha}{\alpha} + c_v\frac{dp}{p} = c_p\frac{d\alpha}{\alpha} + c_v\frac{dp}{p} \qquad (2.33)$$

$$\frac{dq}{T} = d(\ln p^{c_v}\alpha^{c_p}) \qquad (2.34)$$

The expression dq/T is of considerable interest. It represents the change in the entropy of the system per unit mass. It is discussed further in Section 2.12, but for a full discussion of entropy the reader should refer to a standard textbook on thermodynamics.

2.11 ADIABATIC PROCESS

If there is no exchange of heat between a system and its environment the dynamical process by which a perfect gas is heated by compression or cooled by expansion is called an adiabatic process. The gain or loss of heat to the system in an adiabatic process is caused by the loss or gain of energy due to the work which has been done by the environment on the system or by the system on the environment as the system is compressed or as it expands, respectively.

Equation (2.17) may be written

$$dq = du + dw = 0$$

and thus

$$c_v \, dT + p \, d\alpha = 0 \tag{2.35}$$

We may set all the relationships in (2.30) and (2.33) and in (2.31) and (2.34) equal to zero. Thus,

$$\frac{dq}{T} = c_v \frac{dT}{T} + R \frac{d\alpha}{\alpha} = c_p \frac{dT}{T} - R \frac{dp}{p} = c_v \frac{dp}{p} + c_p \frac{d\alpha}{\alpha} = 0 \tag{2.36}$$

and

$$\frac{dq}{T} = d(\ln T^{c_v} \alpha^R) = d(\ln T^{c_p} p^{-R}) = d(\ln p^{c_v} \alpha^{c_p}] = 0 \tag{2.37}$$

Integrating (2.37), we obtain

$$T^{c_v} \alpha^R = \text{constant}$$

$$T^{c_p} p^{-R} = \text{constant} \tag{2.38}$$

$$p^{c_v} \alpha^{c_p} = \text{constant}$$

Equations (2.38) represent the various curves of state of different adiabatic processes, as determined by the initial conditions for p, α, T. Let us define

$$\frac{R}{c_p} = \kappa \tag{2.39}$$

and

$$\frac{c_p}{c_v} = \eta \tag{2.40}$$

Then, from (2.38),

$$T = \text{constant} \, (p^{R/c_p}) = \text{constant} \, (p^{\kappa}) \tag{2.41}$$

and

$$p = \text{constant} \, (\alpha^{-c_p/c_v}) = \text{constant} \, (\alpha^{-\eta})$$

or

$$p\alpha^{\eta} = \text{constant} \tag{2.42}$$

The values of κ and η for dry air are

$$\kappa = 0.286 \qquad \eta = 1.400$$

2.12 POTENTIAL TEMPERATURE

Equation (2.42) represents the family of adiabatic curves on an α, p diagram; when $\eta = \eta_d$ the curves are called dry adiabats. Each point on a dry adiabatic curve represents a temperature. The adiabatic lines are labelled according to the temperature at the point where the curve intersects the 1000 mb isobar. This temperature is called the potential temperature and it is denoted by θ. We may define the potential temperature, therefore, as the temperature assumed by a parcel of air when that parcel is expanded or compressed adiabatically to a pressure of 1000 mb. A value of the potential temperature therefore defines a given adiabatic process, and conversely for any adiabatic process the potential temperature must be constant.

Thus, from (2.41),

$$\frac{T_1}{T_2} = \left(\frac{p_1}{p_2}\right)^\kappa$$

If $p_2 = 1000$ mb and $T_2 = \theta$, the potential temperature, as defined above, is

$$\theta = T \left(\frac{1000}{p\,(\mathrm{mb})}\right)^\kappa \tag{2.43}$$

Equation (2.43) is sometimes called Poisson's equation.

The potential temperature of a parcel of dry air at any pressure and temperature (p, T) can be calculated from (2.43), where $\kappa = \kappa_d$.

If we differentiate (2.43) logarithmically it follows that

$$\frac{d\theta}{\theta} = \frac{dT}{T} - \frac{\kappa\,dp}{p} = \frac{dT}{T} - \frac{R\,dp}{c_p p} \tag{2.44}$$

Multiplying by c_p

$$c_p \frac{d\theta}{\theta} = c_p \frac{dT}{T} - R\frac{dp}{p} \tag{2.45}$$

Comparing this with (2.30) we get

$$\frac{dq}{T} = c_p \frac{d\theta}{\theta} = d(c_p \ln \theta) \tag{2.46}$$

If $dq = 0$ then $d\theta = 0$ and $\theta =$ constant. It is already known by definition that the potential temperature is constant for an adiabatic process.

Worked Example
If the temperature at 500 mb (hPa) is $-20°C$, what is the potential temperature of a parcel of air at that level?

Solution:

$$\theta = T\left(\frac{1000}{p}\right)^\kappa = 253 \times 2^{0.286} = 308.7°\mathrm{K} = 35.7°\mathrm{C}$$

2.13 ENTROPY

It is seen from (2.31), (2.34) and (2.46) that the term dq/T is equal to a total derivative of an expression which defines a state of a gas. The term dq/T is thus called a differential of a function of state of a gas. If

$$\mathrm{d}\phi = \frac{\mathrm{d}q}{T}$$

$$\phi = \int \frac{\mathrm{d}q}{T} + \text{constant} = \int \mathrm{d}(c_p \log \theta) + \text{constant} \qquad (2.47)$$

$$\phi = c_p \log \theta + \text{constant}$$

The quantity ϕ is called specific entropy. It is seen that ϕ increases or decreases as heat is absorbed or removed.

For adiabatic processes $\mathrm{d}q = 0$ and $\phi = $ constant. Consequently adiabatic processes are often called isentropic. Isentropic synoptic charts are sometimes constructed. They are composed of contours of the heights of a selected surface of constant potential temperature above mean sea level (m.s.l.), or of isobars giving the pressures on such a surface. An isentropic process must be reversible.

The quantity entropy is somewhat abstruse and often discussed in physics and in cosmology (e.g. see *A Brief History of Time* by Stephen Hawking). It is sometimes described as a measure of disorder of the universe. However, in meteorology we will not delve into such esoteric concepts but only concern ourselves with the physical meaning of the formulae for $\mathrm{d}q/T$.

Isentropic analysis, which is actually analysis using lines of equal potential temperatures (from equation (2.47)) is particularly useful in tracking air masses as air tends to follow the dry adiabatic lines of constant potential temperature; that is, if the air moves up or down it will cool or warm at the dry adiabatic lapse rate and air parcels may therefore (in theory, assuming no diabatic or non-adiabatic heating) be followed. The amount of warming and/or cooling which occurs as the air moves up or down the dry adiabats may be determined from reading the temperature at the intersection of the lines of constant potential temperature and the isotherms on an aerological or upper air diagram. This will be the subject of the next chapter.

2.14 PROBLEMS

1. What are the units and dimensions of the specific gas constant R_d?
2. Equation (2.6) was derived using Boyle's law and then Charles' law. Derive (2.6) using the opposite sequence, that is Charles' law and then Boyle's law.
3. The pilot of an aircraft flying from Miami to Montreal in winter wishes to know the air density for takeoff at the two terminals. At Miami the surface pressure is 1000 mb (hPa) and the temperature 30°C. At Montreal the surface pressure is 1040 mb (hPa) and the temperature is −20°C. At what pressure over Montreal would the density be the same as at Miami, assuming there was no change of temperature with height at Montreal?
4. Find the amount of work performed in lifting a kilogram of dry air from 1000 hPa to some level where the pressure is p, if the potential temperature is constant throughout the layer.

5. Using the mass–energy equivalence equation $E = mc^2$, how much would 1 kilogram of mass warm the global atmosphere if the heat was uniformly distributed? Assume an average m.s.l. pressure of 1000 hPa.

6. Two parcels of air are at 1000 hPa and 10°C. One undergoes an isothermal process and the other an adiabatic process. What is the ratio of their densities at some higher level p? What is the ratio of the two constants in the relations $p\alpha = $ constant and $p\alpha^\eta = $ constant?

3

THE AEROLOGICAL DIAGRAM

3.1 INTRODUCTION

The thermodynamic or aerological diagram is an indispensable tool to the meteorologist in the analysis of the temperature and humidity structure of the column of air above us. The diagram can be used not only to explain and predict sky conditions on quiet days when the pressure distribution on the weather map is flat, when convection may take major control of the weather, but also on less settled days, as long as an upper air ascent is available within the air mass which is predicted to be overhead at the time it is desired to know the weather and state of sky. The diagram can assist in the prediction and breakup of clouds, and in estimating cloud amounts, cloud bases and tops, and whether showers or thunderstorms are likely to develop or not. It is also particularly useful as a tool to assist in the prediction of the formation and the morning clearance of radiation fog.

The aerological diagram is simply a graph upon which observations of temperature, pressure and moisture content are plotted. Various lines are constructed from theoretical equations and drawn as a permanent backing to the diagram. When curves of the actual temperature and moisture from a given upper air radiosonde or aircraft ascent are plotted and compared with the background lines certain conclusions may be drawn about the vertical structure of the atmosphere. The diagram may look complicated at first because it seems to have so many lines on it, but it is actually quite simple and this will become apparent after some simple plotting exercises are done, perhaps with the current day's upper air ascent obtained from a nearby airport or weather office. Current weather is like today's newspaper: up to date and usually having some points of special interest.

3.2 DIFFERENT KINDS OF DIAGRAM

There are several different kinds of aerological diagrams. The simplest is probably the pressure, volume (p, α) or Clapeyron diagram. It is not a suitable one for practical use since one of the coordinate axes is specific volume. This is not a quantity that can be physically measured by a radiosonde transmitting instrument during its balloon flight into the troposphere, although it can be derived from the equation of state. If one were to draw a (p, α) diagram with an eye to meteorological use it would look like Fig. 3.1. Since potential temperature is conserved in an adiabatic process, the adiabats or isentropes can each be labelled with a particular value of θ.

The figure exhibits several disadvantages of the (p, α) diagram. In the first place, both the isotherms and the adiabats are excessively curved, and in the second place, the angle between them is not very great, considering the wide difference in character between isothermal and adiabatic processes. The diagram has one advantage, however, in that it is an energy diagram in the sense that area on it represents work or energy per unit mass. We will look at the criteria that are needed in the design of an aerological diagram:

(a) The abscissa (x axis) should be temperature, if possible.
(b) The ordinate (y axis) should be a function of pressure, chosen so that it is also approximately a height coordinate.
(c) If some function of pressure is not chosen for the ordinate the resultant isobars should at least not be too curved or crowded together.
(d) The diagram should be an 'equal-area' diagram, in the sense that area is proportional to energy/unit mass, although slow changes in the 'constant' of proportionality over the diagram can be tolerated for the sake of other desirable features.
(e) The adiabatic process for dry air should be well represented, that is the dry adiabats should be fairly straight and nearly perpendicular to the isotherms.

Since temperature and pressure are directly measured in both surface and upper air meteorology it is not surprising to find that most meteorological

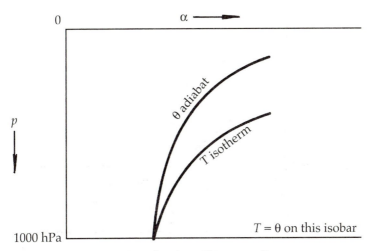

FIGURE 3.1 Schematic form of the (p, a) or Clapeyron diagram.

diagrams are based upon the variables T and p. We will therefore briefly discuss the merits or otherwise of the $(T, -p^\kappa)$, $(T, -p)$ and $(T, -\log p)$ diagrams, where log of course denotes the natural logarithmic function ln.

All of the above satisfy criteria (a) and the first part of (b). However, the $(T, -\log p)$ diagram is the only one which satisfies the requirement that the ordinate be approximately a height scale. Let us now determine which of the diagrams (a)–(e) has the 'equal-area energy property' mentioned in (d).

In the case of isothermal heating, that is the addition of heat energy to a kilogram of dry air, while holding its temperature constant, the last expression on the right hand side of (2.30) reduces to

$$dq = -RT\frac{dp}{p} \tag{3.1}$$

and it is easily shown (problem 2) that the latter can take the alternative forms

$$dq = \frac{c_p}{p^\kappa}T\,d(-p^\kappa)$$

$$dq = \frac{R}{p}T\,d(-p) \tag{3.2}$$

$$dq = RT\,d(-\log p)$$

Since $T\,d(-p^\kappa)$, $T\,d(-p)$ and $T\,d(-\log p)$ all denote elements of area upon diagrams possessing these coordinate axes, it is clear that the scale factors between area and energy per unit mass are respectively proportional to c_p/p^κ, R/p and R. Only the $(T, -\log p)$ diagram has the 'equal-area' property (since R is a constant and p is a variable). However, since $\kappa = 0.286$, the change in scale

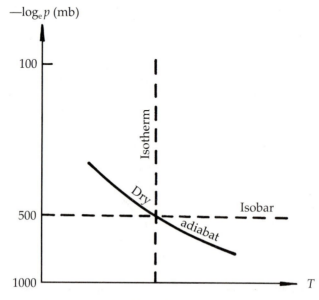

FIGURE 3.2 Schematic form of the $(T, -\log p)$ diagram.

factor with respect to p over the $(T, -p^{\kappa})$ range is sufficiently slow to be tolerable, so the diagram is sometimes used and is known as the Stüve diagram.

Apart from the fact that the Stüve diagram alone has straight adiabats the dry adiabats are not very well represented on any of the above diagrams, mainly because they do not intersect the isotherms nearly at right angles. As an example of this see the schematic version of the $(T, -\log p)$ diagram in Fig. 3.2.

From the above discussion it is clear that the $(T, -\log p)$ diagram satisfies more of the criteria (a)–(e) above than the other two diagrams. For this reason it is quite popular for displaying the results of radiosonde ascents and for computing the heights of various pressure levels. In Europe, the diagram is often known as the Väisälä diagram while in North America it is called the Emagram (energy per unit mass diagram).

3.3 THE SKEW $(T, -\log p)$ DIAGRAM

In Australia the Commonwealth Bureau of Meteorology uses the skew $(T, -\log p)$ diagram or Herlofson diagram. The only difference between it and the $(T, -\log p)$ diagram is that the temperature axis is skewed a further 45° from the $-\log p$ axis to make an angle of 135° with the latter. This transformation allows the dry adiabats to intersect the isotherms at an angle close to 90°. The adiabats are slightly curved. The properties of the skew $(T, -\log p)$ diagram are quite similar to those of the tephigram, the diagram used by the British Meteorological Office.

3.4 THE TEPHIGRAM

If we wished to design a meteorological diagram in which the distinction between the two limiting processes for dry air, namely the adiabatic and isothermal

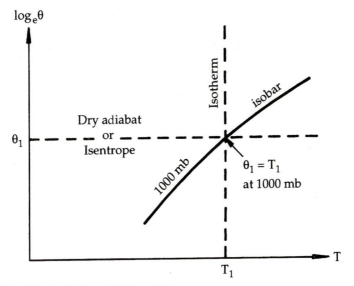

FIGURE 3.3 Schematic form of the tephigram.

processes, is best exhibited, then we would probably choose both the adiabats and the isotherms to be straight lines and we would insist that they intersect everywhere at right angles. Such requirements can in fact be satisfied if we retain temperature as the abscissa of our diagram but take the entropy of 1 kilogram of dry air to be the ordinate. This gives rise to the so-called temperature–entropy diagram, which is widely used by meteorologists, as well as by engineers in non-meteorological contexts. We saw in the last chapter that in equation (2.47) entropy was denoted by the Greek letter ϕ, and that ϕ was proportional to the logarithm of the potential temperature, denoted by θ. In view of these symbols it is not surprising that a diagram constructed with the axes just mentioned should be called a $T-\phi$ gram or tephigram.

Figure 3.3 shows a tephigram in schematic form. Figure 3.4 is a replica of the main working area of the diagram as it is used in practice. The axes have been rotated so that pressure and height are represented as nearly as possible along the vertical. Such an orientation of axes is useful since relative height can be judged at a glance. The pressure lines slope very gently to the right. The isotherms slope upwards at an angle of about 45° to the right and potential temperature lines or dry adiabats at about 45° upwards to the left.

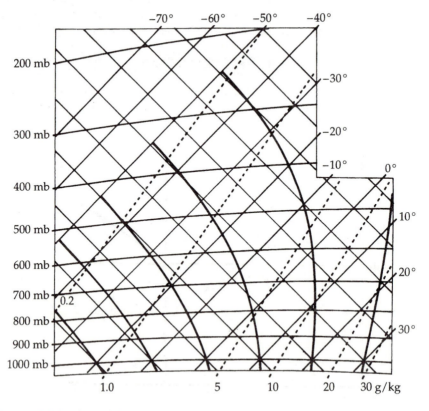

FIGURE 3.4 Replica of the main working area of a tephigram.

The lines of equal pressure are labelled in mb (hPa) at the left hand side of the diagram. The isotherms are labelled in degrees Celsius. The mixing ratio lines are the dotted lines and they are labelled in $g\,kg^{-1}$ of water vapour (see next chapter). The dry adiabatic lines of constant potential temperature which slope upwards from right to left are not labelled but they may be designated by the value of the temperature at the intersection with the 1000 mb (hPa) isobar. Finally, the saturated adiabats are shown by the heavier curved lines. These lines are not labelled but their value may be designated as for the dry adiabatic lines of constant potential temperature. These lines will also be discussed in the next chapter.

It will be noted that the origins of T and ϕ are well off the diagram. Figure 3.3 only shows a small portion of the full theoretical diagram, but it covers the ranges of variables occurring in the lower atmosphere.

From (2.45) and (2.46)

$$d\left(c_p \log \theta\right) = c_p \frac{dT}{T} - R\frac{dp}{p}$$

The equation of an isobar is then

$$d\left(c_p \log \theta\right) = c_p \frac{dT}{T}$$

$$c_p \log \theta = c_p \log T + \text{constant} \tag{3.3}$$

or

$$\log \frac{\theta}{T} = \text{constant}$$

The equation of a dry adiabat is already given in (2.43).

3.5 WORK AND ENERGY ON THE TEPHIGRAM

It is clear from (2.15) that the work performed in a cyclic process can be represented on an (p, α) diagram by the area enclosed by the path of the process.

From (2.17)

$$\oint dq = \oint du + \oint dw$$

$$= \oint c_v\, dT + \oint dw$$

but

$$\oint c_v\, dT = 0 \quad \text{since } c_v \oint dT = 0$$

because there is no change in temperature around a closed path. The integral around a closed path of an exact differential representing a function of state must be zero. The reader should refer to a textbook on calculus for a rigorous proof of this theorem.

Thus

$$\oint dq = \oint dw \tag{3.4}$$

and from (2.46)

$$\oint dq = \oint T d(c_p \log \theta)$$

$$\oint dw = \oint T d(c_p \log \theta)$$

Thus

$$w = \oint T d(c_p \log \theta) \tag{3.5}$$

and $w = A$ where A is an area on the tephigram, since $T d(c_p \log \theta)$ is obviously an element of area on the tephigram. Thus the work performed by a cyclic process can be represented by the area enclosed by the path of the process on a tephigram. The application of this concept will be discussed in greater detail in a later chapter. It has an important bearing on the use of the tephigram and other diagrams since it enables a calculation to be made of the potential thermo-dynamic energy of the air above us. This has come to be known by the acronym CAPE (Convective Available Potential Energy). It is represented by an area on the diagram produced by a cyclic process like that described by equation (3.5). It is this particular feature of the diagram that enables the weather forecaster to predict showers, thunderstorms, the breakup of radiation fog and low stratus cloud. But to do this we must first study the properties of water vapour and the behaviour of moist air, that is the mixture of dry air and water vapour. We will discuss this subject in the next chapter.

3.6 PROBLEMS

1. Which equations enable us to plot values of T and θ all over a Clapeyron diagram and hence to construct isopleths of temperature and potential temperature on the diagram?
2. Show that $dQ = (c_p/p^\kappa) T d(-p^\kappa)$ reduces to $dQ = -RT (dp/p)$.
3. Why are the isobars on a tephigram not straight lines?
4. The station-level pressure at Adelaide Airport is 1020 mb at the same time as a barometer nearby at Mount Lofty Lookout reads 950 mb.

    (a) Assuming that dry air, which has a temperature of 26.5°C at the Airport, flows rapidly from there to the Lookout without undergoing any diabatic (i.e. non-adiabatic) heating, find the temperature which it would have at the Lookout.

    (b) Assuming that this air then flows adiabatically into a valley where the pressure at the valley floor is 980 mb, what would you expect the temperature in the valley to be?

5. It is perhaps not immediately obvious that the skew $(T, -\log_e p)$ diagram is an 'equal-area' energy diagram and a rather neat geometric proof of this is as follows. Take a tephigram and dissect it into a large number of infinite-simal isothermal strips. Slide these strips parallel to one another until the isobars are straight lines (in the limit of finer and finer dissections). Since

the 'equal-area' energy property of the original tephigram is preserved under this transformation and the latter transforms a tephigram into a skew $(T, -\log_e p)$ diagram, it follows that the skew $(T, -\log_e p)$ diagram is itself an 'equal-area' energy diagram.

6. Plot the following points on an aerological diagram:

    - $p = 1000\,\text{mb}$, $T = 15°\text{C}$
    - $p = 500\,\text{mb}$, $T = -21.2°\text{C}$
    - $p = 220\,\text{mb}$, $T = -56.5°\text{C}$.

7. Obtain a current (or recent day's) upper air ascent from your nearest major airport or weather office. Plot the temperature and pressure for each pair of coupled readings. Describe the curve.

4

THE THERMODYNAMICS OF MOIST AIR

In the previous chapter our considerations have been confined to perfect gases and subsequently to a mixture of perfect gases which is known as dry air. In the atmosphere air is never completely dry. It may be relatively dry over deserts and at very high altitudes but there is always some water vapour in it. The continuous evaporation of water into the atmosphere from the vast oceans and inland waters as well as from the ground and from vegetation is the source of all the clouds and varied forms of condensation and precipitation which go to make up the weather and climate of our globe.

Water vapour is a gas which behaves in the same way as other gases. It obeys the various laws which we have already discussed. It is a constituent of the earth's atmosphere and obeys Dalton's law in the same way as the other gaseous constituents. However, a substance behaves according to the various gas laws just as long as it is in a gaseous phase and does not liquefy or solidify.

Water substance, however, does liquefy and solidify within a range of temperatures which commonly occur.

Now only a certain amount of water can exist in a given volume in the gaseous phase. This amount varies according to the temperature. Thus, if water is injected into a vacuum of given volume it will at first evaporate and exert a vapour pressure e. After a while, if further water is injected, the additional water will not evaporate and the vapour pressure e will remain constant. Any further water which may be injected will remain in the liquid state. The space is then said to be saturated and the pressure e of the vapour at the point when it no longer increases, that is when evaporation ceases, is the saturation vapour pressure. This saturation vapour pressure e_s varies according to the temperature. It is, in fact, a function of the temperature only and increases with the temperature.

The reason is that at higher temperatures the molecules of water at the surface

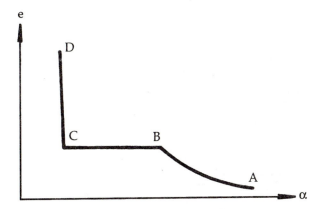

FIGURE 4.1 Curve of state of water vapour at 300 K.

of the liquid are moving more rapidly and more of them overcome the attraction exerted on them by other molecules in the liquid state. These then escape into the less dense gaseous state.

A vapour whose pressure is less than the saturation vapour pressure is said to be unsaturated and if brought into contact with its own liquid, the latter will evaporate until it has all gone or until the vapour has reached the saturation pressure, or in other words, until the space occupied by the vapour has become saturated.

Suppose now we have a cylinder containing a perfectly fitted piston. Let the cylinder be filled with vapour and let it be unsaturated. Let the system remain at a constant temperature T, say 300 K. Now compress the gas by means of the piston. The path of state of the water vapour will then be observed to follow a curve as shown in Fig. 4.1. The first effect of an increase in pressure will be a decrease of volume. This occurs along the path AB. At B the water vapour reaches the saturation vapour pressure and it starts to condense; any further compression by the piston will cause the remaining water vapour to condense to liquid water. The volume in the cylinder thus decreases without any further change in the vapour pressure e_s. C represents the stage of the process where all the vapour has condensed to water, which is virtually incompressible, so that further compression by the piston does not change the volume significantly.

The above curve of state will be observed if the system is at about 300 K. Different curves will be observed for other temperatures. We may draw a diagram showing the state of water in all its phases, solid, liquid and gas at all temperatures. Such a diagram is shown in Fig. 4.2. The diagram is divided into regions showing the states of vapour, combination of water and vapour, combination of ice and vapour and water only. At very high temperatures the water vapour never condenses, however great the pressure, and the curve of state is similar to that of a perfect gas. This occurs for isotherms $T > T_c$. The value T_c is the temperature at which the vapour stage touches the water and vapour stage. T_c is called the critical temperature. The critical pressure e_c is the highest pressure at which liquid water and water vapour can exist in co-equilibrium.

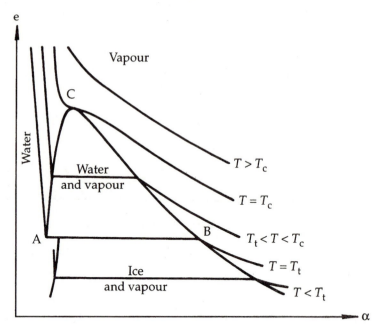

FIGURE 4.2 State of water in its three phases.

The critical specific volume α_c is the value of α observed at e_c, T_c. The state e_c, α_c, T_c is called the critical state. The values are

- $e_c = 221\,000\,\text{hPa} = 218$ atmospheres
- $T_c = 647\,\text{K}$
- $\alpha_c = 3.1 \times 10^{-3}\,\text{m}^3\,\text{kg}^{-1}$.

At lower temperatures a state of coexistence between liquid water and saturated water vapour occurs within the region bounded by ABC. Further compression causes all the vapour to condense and the total liquid phase occurs to the left of AC. When the temperature is reduced sufficiently a value is reached where the water freezes. The temperature at which liquid water, ice and water vapour exist in coequilibrium is called the triple state temperature and this state of the water substance is called the triple state, or sometimes the triple point. It occurs along the line AB. The triple state temperature is $T_t - t_0$. t_0 is the temperature of a substance which is brought into equilibrium with a mixture of ice and pure water at a pressure of 1 atmosphere (1013.2 mb). This temperature is 0°C. Also we may define t_{100} as the temperature of a substance which is brought into thermal equilibrium with steam immediately over water boiling at a pressure of one atmosphere (1013.2 hPa). This temperature is 100°C. The values of e, T, and α at the triple state are

- $e_t = 6.11\,\text{hPa}$
- $T = 0.0075°C$
- $\alpha_t = 1.0 \times 10^{-3}\,\text{m}^3\,\text{kg}^{-1}$ for water

- $\alpha_t = 1.091 \times 10^{-3} \, \text{m}^3 \, \text{kg}^{-1}$ for ice
- $\alpha_t = 206.2 \, \text{m}^3 \, \text{kg}^{-1}$ for water vapour.

At temperatures lower than T_t a state of ice and vapour occurs. At very great pressures the ice is converted into water, which freezes again when the pressure is released. This phenomenon is called regelation and accounts for the flow of glaciers and our ability to ski and skate.

In the above discussion the subject of the supercooling of water, that is the cooling of water below T_t, has been omitted. This will be mentioned later.

4.2 EQUATION OF STATE FOR WATER VAPOUR

We may now write the equation of state of a perfect gas for water vapour, following the same form as (2.8) for dry air:

$$e\alpha_v = R_v T \tag{4.1}$$

where e, the pressure of the water vapour, replaces p and

$$R_v = \frac{R^*}{m_v} = 461 \, \text{J} \, \text{kg}^{-1} \, \text{K}^{-1}$$

Worked Example
What is the density of water vapour at 25°C and vapour pressure 25 hPa? How does this compare with dry air at the same pressure and temperature? Could you have anticipated this result?

Solution:
We start with equation (4.1) in the form

$$\rho = \frac{e}{R_v T}$$

$$= \frac{2500}{461 \times 298} \, \text{kg} \, \text{m}^{-3}$$

$$= 18.2 \, \text{g} \, \text{m}^{-3}$$

For dry air under the same conditions we have

$$\rho = \frac{2500}{287 \times 298 \, \text{kg} \, \text{m}^{-3}} = 29.2 \, \text{g} \, \text{m}^{-3}$$

We note that water vapour is about 0.62 times as heavy as dry air. This could have been anticipated from the ratio of the molecular weights of water and dry air.

Now

$$m_v R_v = m_d R_d \quad \text{since } mR = R^*$$

and so

$$R_v = \frac{m_d}{m_v} R_d = \frac{1}{\epsilon} R_d$$

where ϵ is the ratio of the molecular weight of water vapour to that of dry air and equals 0.622.

Equation (4.1) may be rewritten in the form

$$e\alpha_v = \frac{R_d}{\epsilon} T = R_v T \tag{4.2}$$

We may compare (4.2) with (2.8).

4.3 SPECIFIC HEATS OF WATER SUBSTANCE

The specific heats of water and ice may be considered constant for atmospheric problems. The values in m.k.s. mechanical units are:

- $c_w = 4185 \, \text{J kg}^{-1} \, \text{K}^{-1}$
- $c_i = 2060 \, \text{J kg}^{-1} \, \text{K}^{-1}$.

The specific heats of water vapour at constant pressure and constant volume are, respectively,

- $c_{pv} = 1911 \, \text{J kg}^{-1} \, \text{K}^{-1}$
- $c_{vv} = 1450 \, \text{J kg}^{-1} \, \text{K}^{-1}$.

These values may be up to 2% in error since water vapour does not always behave as a perfect gas. However, $c_{pv} - c_{vv} = R_v = 461 \, \text{J kg}^{-1} \, \text{K}^{-1}$ which agrees with the value previously given.

4.4 CHANGE OF PHASE

We have seen that it is possible for saturated vapour and its liquid to exist together side by side in equilibrium through a range of pressures and temperatures, and that the vapour pressure at which water and water vapour exist side by side is the saturation vapour pressure, e_s. We will now consider what happens when water substance changes from one physical state or phase to another. Such a change will involve a change in the total heat content of the system.

Let the two phases in equilibrium be 1 and 2. Then, from (2.18) and Section 2.12,

$$\int_1^2 dq = \int_1^2 T \, d\phi = \int_1^2 du + \int_1^2 e_s \, d\alpha \tag{4.3}$$

The first integral represents the total amount of heat absorbed by unit mass of the substance in phase 1 to transform it to phase 2. It is known as the latent heat of transformation and will be denoted by $L_{1,2}$. $L_{1,2} = -L_{2,1}$; that is, it also equals the amount of heat released by the substance during the transformation from phase 2 to phase 1. By convention the change of heat dq is positive if the heat is absorbed by the system and negative if it is released. If we integrate (4.3), holding pressure and temperature constant during the transformation we obtain

$$L_{1,2} = T(\phi_2 - \phi_1) = u_2 - u_1 + e_s(\alpha_2 - \alpha_1) \tag{4.4}$$

$L_{1,2}$ varies with the temperature and has a different value for each of the three phase transformations. These are

$L_{w,v}$, liquid water $\leftrightarrow$ vapour   Latent heat of evaporation

$L_{i,v}$, ice $\leftrightarrow$ vapour   Latent heat of sublimation

$L_{i,w}$, ice $\leftrightarrow$ liquid water   Latent heat of melting

The values are

- $L_{w,v} = 2.500 \times 10^6\,\text{J}\,\text{kg}^{-1}$
- $L_{i,v} = 2.834 \times 10^6\,\text{J}\,\text{kg}^{-1}$
- $L_{i,w} = 0.334 \times 10^6\,\text{J}\,\text{kg}^{-1}$.

Actually these latent heats do depend upon temperature, but they are often taken to be constant within the atmospheric range of temperature. The three cases stated above are all positive since heat must be absorbed to evaporate, to sublimate or to melt water substance. When water substance condenses, or freezes, heat is liberated and the values for the latent heats are negative.

4.5 VARIATION OF LATENT HEAT WITH TEMPERATURE

The variation of latent heat with temperature is relatively small and may be obtained as follows: we will take the case of $L_{w,v}$. From (4.4)

$$L_{w,v} = u_v - u_w + e_s(\alpha_v - \alpha_w)$$

$$= e_s\alpha_v - e_s\alpha_w + u_v - u_w$$

$\alpha_w = 1$ and may be neglected in comparison with α_v.

Now, $e_s\alpha_v = R_vT$ from the equation of state; differentiating,

$$dL_{w,v} = R_v\,dT + du_v - du_w = R_v\,dT + (c_{vv} - c_w)\,dT$$

$$\frac{dL_{w,v}}{dT} = R_v + c_{vv} - c_w = c_{pv} - c_w \tag{4.5}$$

$$\frac{dL_{w,v}}{dT} = 1911 - 4185 = -2274\,\text{J}\,\text{kg}^{-1}\,\text{K}^{-1}$$

It is noted that 2274 is much smaller than 2.5 million, the value of L_{wv}. However, it may be shown that over an atmospheric temperature range from $-40°C$ to $+40°C$ the range undergoes a change of about 7%. Whilst the change is by no means negligible for very accurate work, it does suggest that we can take L to be constant in order to obtain a first approximate solution of the differential equation which we will now derive.

4.6 CLAPEYRON'S EQUATION

The equation called the Clausius–Clapeyron equation is one of the most important equations in the thermodynamic subheading of meteorology. It shows the physical relation between the saturation vapour pressure and the temperature. It is a little difficult to follow so we will develop it step by step.

The first law of thermodynamics for water substance may be written in the form

$$dQ = T\,d\phi = du + e_s\,d\alpha \tag{4.6}$$

This follows from equation (2.18) where p has been replaced by e_s, the saturation vapour pressure of water vapour; du denotes the change in internal energy and $d\alpha$ is the change in specific volume which occurs when an amount of heat dq is imparted to 1 kilogram of water substance. Let us suppose that the suffixes 1 and 2 represent any two phases of water substance which are in equilibrium with one another at a temperature T, for example water vapour in contact with a water surface. Then, if $L_{1,2}$ denotes the latent heat associated with the change from phase 1 to phase 2, we have, from (4.6),

$$L_{1,2} = dQ = du + e_s\,d\alpha \tag{4.7}$$

and since the pressure e_s remains constant whilst the specific volume changes during a change of phase, it follows that

$$L_{1,2} = T(\phi_2 - \phi_1) = u_2 - u_1 + e_s(\alpha_2 - \alpha_1) \tag{4.8}$$

In classical thermodynamics, the combination of variables

$$G = u + e_s\alpha - T\phi \tag{4.9}$$

is called the Gibbs function and it is only a function of state of the water substance. From (4.9) it is clear that

$$u_1 + e_s\alpha_1 - T\phi_1 = u_2 + e_s\alpha_2 - T\phi_2 \quad \leftrightarrow \quad G_1 = G_2 \tag{4.10}$$

during the isothermal change of phase under consideration.

We must include a word of caution here in dealing with this thermodynamic principle. By phase we mean the physical state of a substance, that is gas, liquid or solid. Where H_2O substance is concerned, we commonly call the three phases water vapour, water and ice. In addition to the physical state of a substance we also have its thermodynamic state. This is the state we mean when we talk about the equation of state which was introduced very early in this text (Section 2.2). It is the state of a phase which is known if two variables of the three given in the equation of state are known. Now that this distinction between state and phase is defined we may continue with the derivation of our equation.

Suppose that $G + dG$ represents the Gibbs function for a neighbouring state $(T + dT, e_s + de_s)$ of water substance. Then, for an isothermal change of phase in this neighbouring state we would have $G_1 + dG_1 = G_2 + dG_2$, according to (4.10). Subtraction of the two equations above then tells us that $dG_1 = dG_2$. Now taking the differential of (4.10), we have that

$$dG = du + e_s\,d\alpha - T\,d\phi + \alpha\,de_s - \phi\,dT \tag{4.11}$$

where it must be emphasized that the derivative d( ) is associated with the change to the neighbouring thermodynamic state rather than with the change of phase. Since the differential form $T\,d\phi = du + e_s\,d\alpha$ of the first law of thermodynamics is as valid for changes of state as it is for changes of phase, it may be used to reduce (4.11) to

$$dG = \alpha\,de_s - \phi\,dT \tag{4.12}$$

Since $dG_1 = dG_2$ we have

$$\alpha_1 \, de_s - \phi_1 \, dT = \alpha_2 \, de_s - \phi_2 \, dT$$

which can be written in the form

$$\frac{de_s}{dT} = \frac{(\phi_2 - \phi_1)}{(\alpha_2 - \alpha_1)} \qquad (4.13)$$

Since we are not particularly interested in entropy and $\phi_2 - \phi_1 = L_{1,2}/T$ as defined by (4.8), a more useful form of (4.13) is

$$\frac{de_s}{dT} = \frac{L_{1,2}}{T(\alpha_2 - \alpha_1)} \qquad (4.14)$$

which is the famous Clausius-Clapeyron equation which gives us the saturation vapour pressure as a function of temperature. Applying the above to all of the possible changes of phase of water substance, we find that for equilibrium between

(a)   water and water vapour

$$\frac{de_{sw}}{dT} = \frac{L_{w,v}}{T(\alpha_v - \alpha_w)} \qquad (4.15)$$

This equation covers the processes of evaporation and condensation. The subscripts w, v refer to the liquid and gaseous phases of water substance respectively.

(b)   ice and water vapour

$$\frac{de_{si}}{dT} = \frac{L_{i,v}}{T(\alpha_v - \alpha_i)} \qquad (4.16)$$

This equation covers the process of sublimation, the direct transformation from the solid to the gaseous phase and vice versa. It occurs when the temperature is less than the triple point although the process (a) may also occur at temperatures below the triple point if water droplets are cooled below 0°C without freezing to ice.

(c)   ice and water

$$\frac{de_s}{dT} = \frac{L_{i,w}}{T(\alpha_w - \alpha_i)} \qquad (4.17)$$

This equation covers melting, freezing and regelation.

We are now in a position to investigate quantitatively the dependence of saturation vapour pressure upon temperature.

Since $\alpha_v = 10^3 \, m^3 \, kg^{-1}$ approximately and α_w is $10^{-3} \, m^3 \, kg^{-1}$ it is clear that $\alpha_v \gg \alpha_w$ and hence to a very good degree of approximation the equation (4.15) reduces to

$$\frac{de_{sw}}{dT} = \frac{L_{w,v}}{T\alpha_v}$$

which from (4.2) may be written in the form

$$\frac{de_{sw}}{dT} = \left(\frac{\epsilon L_{w,v}}{R_d}\right)\left(\frac{e_{sw}}{T^2}\right) \tag{4.18}$$

Assuming $L_{w,v}$ to be independent of temperature, as argued in Section 4.5, the differential equation separates immediately to

$$\frac{de_{sw}}{e_{sw}} = \left(\frac{\epsilon L_{w,v}}{R_d}\right)\frac{dT}{T^2} \tag{4.19}$$

which, upon integration, yields

$$\log e_{sw} = -\left(\frac{\epsilon L_{w,v}}{R_d T}\right) + \text{constant} \tag{4.20}$$

In order to evaluate the constant of integration we make use of the triple point values already given. We may then express the Clausius–Clapeyron equation in the forms

$$e_{sw}(\text{mb}) = 6.11 \exp\left[\left(\frac{\epsilon L_{w,v}}{R_d}\right)\left(\frac{1}{273} - \frac{1}{T}\right)\right] \tag{4.21}$$

$$e_{si}(\text{mb}) = 6.11 \exp\left[\left(\frac{\epsilon L_{i,v}}{R_d}\right)\left(\frac{1}{273} - \frac{1}{T}\right)\right] \tag{4.22}$$

A schematic diagram of the solutions of equations (4.21) and (4.22) is shown in Fig. 4.3. The dashed line represents the equilibrium which can exist between

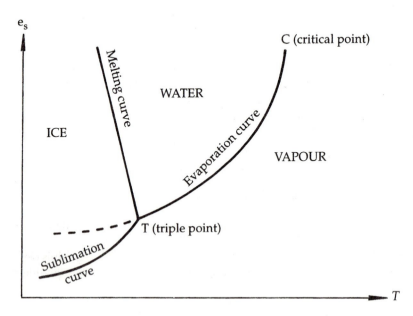

FIGURE 4.3 Evaporation melting and sublimation curves.

supercooled water and water vapour. The melting curve represents the equilibrium between ice and water. Although of geological importance, as for example in connection with the slow flow or creep of glaciers, it is not of great interest to meteorologists (mainly because e_s no longer represents vapour pressure). However, skaters and skiers will be aware of the need to have a slippery surface, and this results in part from the process of regelation, the melting of ice under pressure, creating a film of liquid water. A more exact diagram between the ranges −80°C to 60°C is given in Fig. 4.4. We note that equation (4.20) may be plotted as a straight line on a graph with one axis scaled in $\ln e_s$ intervals and the other axis in $10^3/\mathrm{K}$ intervals. Such a linear relation is shown in Fig. 4.5.

Worked Example
Compute the saturation vapour pressure at 20°C from Clapeyron's equation. Assume $L = 2.5 \times 10^6 \,\mathrm{J\,kg^{-1}}$.

Solution:
We know that e_s is 6.11 hPa at 273 K. Then

$$\log\left(\frac{e_s}{6.11}\right) = \frac{(0.622L \times 20)}{(287 \times 273 \times 293)}$$

$$e_s = 23.7\,\mathrm{hPa}$$

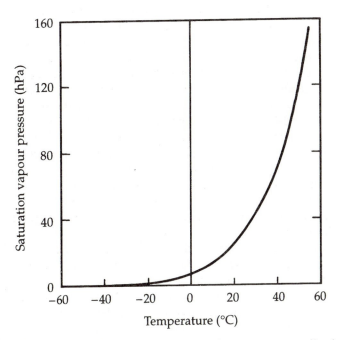

FIGURE 4.4 Curve of saturation vapour pressure against temperature allowing for the change of latent heat with temperature.

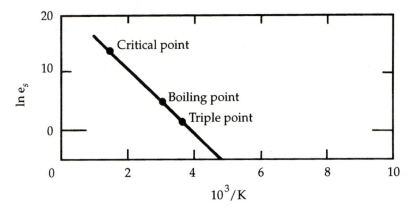

FIGURE 4.5 Evaporation curve on a logarithmic scale with latent heat assumed constant.

4.7 CLAPEYRON AND GLOBAL WARMING

We see that the higher the temperature the higher the saturation vapour pressure which is in equilibrium with liquid water at that temperature. At ordinary temperatures the increase in saturation vapour pressure is about 6% for a 1 K rise in temperature. Now there is no definite mathematical relation between the saturation vapour pressure in equilibrium with a water surface at a given temperature, and the actual vapour pressure in the air at some arbitrary distance above it. Thus, we may know the saturation vapour pressure at the interface of the ocean and atmosphere but this does not mean that we know the exact vapour pressure at, say, 100 metres above the interface, and we would know even less about the vapour pressure at higher levels. This is because the water vapour content of different layers of the atmosphere depends on a number of factors, such as the temperature structure of the layers, whether the air in the different layers is rising or descending (subsiding), and on the rate of evaporation of water from the ocean surface, which is dependent on several variables such as the intensity of solar insolation, the wind strength and the existing moisture content of the air which may have been advected from a dry continent. We have also noted that the increase of saturation vapour pressure with temperature is not linear but exponential. We sometimes use words such as 'sultry' or 'close' to describe the unpleasant feeling induced by the combination of high temperature and high humidity.

Numerical modellers have estimated that for every degree rise in temperature resulting from a greenhouse gas other than water vapour, the temperature will increase a further 0.7°C as a result of the additional water vapour held by the atmosphere because of its higher temperature. That is, if the temperature of the surface of the oceans and of the atmosphere is increased by 1 K, the saturation vapour pressure will have increased and the atmosphere will be capable of taking up at least some extra water from the ocean interface. Since water vapour is a strong greenhouse gas and there will be more of it, the net effect will be to increase the temperature by a further amount. This is what is known as a positive feedback. It is a somewhat unstable process in which any impulse tends to amplify. In the case of a negative feedback the opposite occurs and the system will tend to return to its original state before it was perturbed. We also note from the more accurate form of Clapeyron's equation in which the value of $L_{w,v}$ is allowed to vary in accordance with equation (4.5) that the value of the latent heat of transformation decreases with temperature so that less heat is required to evaporate water at higher than at lower temperatures. This is an additional positive feedback, although a small one.

A word of warning should be given here. The conclusion that an increase in temperature of the atmosphere leads to a positive feedback because of an increase in water vapour is one that has been accepted by the majority of atmospheric scientists. However, there are a minority who believe that the latter conclusion is too simplistic, that there are complex compensations. This somewhat radical opinion suggests that an increase in water vapour in the atmosphere will result in more clouds and that, in consequence, the albedo or reflective property of the planet will be increased, thus decreasing the radiative equilibrium temperature. Such a response would be a negative feedback, a result which would tend to neutralize any increase, at least in the time scale encompassed by global warming predictions.

Certainly, Clapeyron's equation has taken on immense new importance as a result of global concern about the 'greenhouse effect'.

4.8 SUPERCOOLED WATER

Water normally freezes at about 0°C. However, when it is in the form of tiny droplets the surface tension inhibits the freezing process. Water droplets in clouds are very small and may not freeze spontaneously. However, if they come into contact with a solid surface, such as, for example, the wing of an aircraft, the droplets will freeze. Aircraft icing used to present a great danger to aircraft, particularly in winter, in days when aircraft flew at much lower heights than today. Water which remains in the liquid state below 0°C is known as supercooled water. This is an unstable state, and should supercooled water come into contact with ice or with certain other substances in the finely divided state (called *freezing nuclei*) solidification takes place very rapidly. Cloud droplets formed by condensation above 0°C will normally assume this supercooled state on cooling below 0°C. Most cloud elements are still liquid at −10°C and water droplets may be found down to −40°C. When ice crystals are injected into such a cloud of supercooled water drops at a fixed temperature the system is no longer in equilibrium. The vapour is saturated with respect to the water drops and supersaturated with respect to the ice particles. The result is condensation on the ice particles. This reduces the vapour pressure in the air to below the saturation vapour pressure over water. Water therefore evaporates. The net result of the two processes is the growth of ice crystals at the expense of water drops. The process goes on most rapidly near −12°C where the saturation vapour pressure over water most greatly exceeds that over ice.

The Bergeron–Findeisen theory of the mechanism of precipitation is based on the above principle. When a relatively small number of ice crystals are present in a cloud of supercooled water droplets the ice crystals grow to such a size by evaporation from the water droplets that they can no longer remain suspended in the air and start falling. The formation of the ice crystals depends upon the existence of suitable condensation nuclei which are usually present in the atmosphere in large concentrations; however, recent observations suggest that ice-nucleating agents are not as efficient as the nuclei which initiate the condensation of small water droplets. The ice nuclei require a considerable degree of supersaturation of vapour with respect to ice before they become effective in creating ice crystals.

It has been observed in recent years, particularly in tropical regions, that rain can fall from clouds which are above the freezing point. A process of coalescence has been postulated to explain this kind of rain, which is sometimes called 'warm' rain, in contrast to the so-called 'cold' rain formed by the ice crystal theory.

In the 1960s and 1970s a considerable amount of publicity was given to the stimulation of the rain-producing mechanism. Experiments in this activity, known as artificial precipitation, or more popularly, 'rainmaking', were carried out by national meteorological services and by commercial organizations. A favourite method was to 'seed' clouds with silver iodide crystals, dropped from aircraft. The silver iodide acted as a surrogate for ice crystals, as a catalyst to prime the evaporation of supercooled liquid droplets onto the crystals, where they would freeze into ice crystals and subsequently fall out, melt on the way down to the surface and turn to raindrops. It was hoped that these operations would increase precipitation in arid zone countries and in periods of prolonged dry spells or droughts. However, clearly rain could not be stimulated unless the right kinds of clouds were present at the time of seeding. It was often difficult to make useful comparisons between a selected control station and an operational result. It was believed that the technique worked best in trying to increase snowcover over mountains during the winter. However, these operations and experiments could never be regarded as highly successful and interest in them has waned during more recent years.

4.9 MOIST AIR

We have so far discussed only one parameter which is a measure of the water vapour in the air, that is, the vapour pressure (and the saturation vapour pressure). We have seen that the constituent gases of the atmosphere do not affect these values, which are the same as they would be if only water vapour were present. Thus, water vapour is a constituent of the atmosphere along with the other gases; its presence in the atmosphere forms a mixture which we shall call 'moist air'. Although saturation vapour pressure is useful in deriving Clapeyron's equation, we also need a more practical measurement which tells us the actual amount (mass) of water vapour present. We shall therefore now introduce new parameters:

1. The humidity mixing ratio (x) which is defined to be the mass of water vapour per unit mass of dry air.
2. The specific humidity (s) which is defined to be the mass of water vapour per unit mass of moist air.
3. The relative humidity (RH) which is defined as the ratio of the observed vapour pressure to the saturation vapour pressure at the observed temperature, that is e/e_s.

Now, if M' grams of water vapour are mixed with M_d grams of dry air to give $M = M' + M_d$ grams of moist air, then $x = M'/M_d$ and $s = M'/M$. Since x and s are usually less than 0.05, we use the SI unit of g kg^{-1} when specifying their values. Although x and s are usually numerically close, they are still theoretically distinct. From the above definitions it follows that $s = x/(1+x)$ and $x = s/(1-s)$. We will normally use the mixing ratio rather than the specific humidity. Suppose that $(1+x)$ kilograms of moist air occupy a volume V and exert a pressure p. If p_d and e denote the partial pressures of the dry air and water vapour respectively, then

$$p = p_d + e \qquad (4.23)$$

from Dalton's law of partial pressures.

From the perfect gas law

$$p_d \alpha_d = R_d T$$

and also

$$e \alpha_v = R_v T$$

and on dividing the latter by the former we get

$$\frac{e}{p_d} = \frac{x R_v}{R_d}$$

or

$$\frac{e}{(p-e)} = \frac{x}{\epsilon}$$

and

$$x = \frac{\epsilon e}{(p-e)} \approx \frac{\epsilon e}{p} \qquad (4.24)$$

Since (4.24) is true in the case where x and e take their saturation values x_s and e_s, it follows that

$$x_s = \frac{\epsilon e_s}{(p-e_s)} \approx \frac{\epsilon e_s}{p}$$

Taken together, Clapeyron's equation and equation (4.9), recalling that ϕ is a function of temperature and pressure, tell us that the saturation mixing ratio is a known function of temperature and pressure, that is $x_s = x_s(T, p)$. It is the knowledge of this function which enables the isopleths of humidity mixing ratio to be constructed on aerological diagrams. These lines may be identified on Fig. 3.3. They are the lightly dashed lines that slope upwards from left to right and they are labelled in grams per kilogram along the lower boundary of the diagram.

(Do not confuse with the more heavily marked dashed line for the freezing temperature $(0°C)$.)

4.10 THE VIRTUAL TEMPERATURE

We now introduce a corrected form of temperature which can be used when applying 'dry-air' theory to moist-air problems. Although the correction is not large and may not be needed in much of the work which is to follow, nevertheless, bearing in mind the importance of accurate calculations in climate change models, it is necessary to know in what circumstances to use the corrected value.

We have already seen from (2.12) that for unit mass of moist air $1R = M_d R_d + M_v R_v$ where R is the gas constant for moist air, and R_v is the gas constant for water vapour. Then

$$R = (1 - s)R_d + sR_v$$

and

$$R = (1 - s)R_d + \frac{sR_d}{\epsilon} = R_d\left[1 + s\left(\frac{1}{\epsilon} - 1\right)\right]$$

evaluating $\epsilon^{-1} - 1$ and using the approximate relation $s = x$ we get

$$R = (1 + 0.61x)R_d \tag{4.25}$$

Inserting the value of R above in the equation of state for moist air we get

$$p\alpha = (1 + 0.61x)R_d T \tag{4.26}$$

Now we define the quantity $(1 + 0.61x)T$ as T^*, the virtual temperature of the moist air so that $p\alpha = R_d T^*$. Obviously, T^* is the temperature of dry air having the same pressure and specific volume as the moist air. For many purposes moist air may then be treated as dry air of temperature T^*.

4.11 SPECIFIC HEATS OF MOIST AIR

Let the quantity of heat dq be applied to a kilogram of moist air. The moist air is then heated from temperature T to temperature $T + dT$. Then $1 \cdot dq = M_d dq_d + M_v dq_v$ where dq_d is the amount of heat received by the dry air per kilogram of dry air and dq_v is the amount of heat received by the water vapour per kilogram of water vapour.

Dividing both sides by dT and substituting in terms of specific humidity, s, it is seen that

$$\frac{dq}{dT} = (1 - s)\frac{dq_d}{dT} + s\frac{dq_v}{dT}$$

or

$$c_p = (1 - s)c_{pd} + sc_{pv} = c_{pd}\left[1 + s\left(\frac{c_{pv}}{c_{pd}} - 1\right)\right]$$

Evaluating c_{pv}/c_{pd} we get

$$c_p = (1 + 0.90s)c_{pd} = (1 + 0.90x)c_{pd} \tag{4.27}$$

and similarly

$$c_v = (1 + 1.02s)c_{vd} = (1 + 1.02x)c_{vd} \qquad (4.28)$$

We thus see how the specific heat of moist air can be expressed in terms of the specific heat of dry air and the amount of water vapour contained in the air.

4.12 ADIABATIC PROCESS OF UNSATURATED AIR

The adiabatic process for unsaturated air is a special case of the adiabatic process of any perfect gas. Thus from (2.39) and (2.40) $\kappa = R/c_p$ and $\eta = c_p/c_v$ for moist air.

From (4.25) and (4.27)

$$\kappa = \frac{(1 + 0.61x)R_d}{(1 + 0.90x)c_{pd}} = \frac{(1 + 0.61x)\kappa_d}{(1 + 0.90x)}$$

Since $x \ll 1$ we can expand the denominator as a geometric series giving

$$\kappa = (1 - 0.29x)\kappa_d \qquad (4.29)$$

and similarly

$$\eta = (1 - 0.12x)\eta_d \qquad (4.30)$$

It can be shown that Poisson's constant κ for moist air will be lowered from its dry-air value of 0.286 to a minimum of 0.283 for saturated air at the highest temperature likely to be found in the atmosphere.

The equation (2.43) $\theta = T(1000/p)^\kappa$ defines a family of unsaturated adiabats each one of which is a function of an initial p, T and of a given value of κ. Since κ varies with x there would be a different unsaturated adiabat for each value of x for similar initial pressure and temperature values. However, κ departs from κ_d so little that unsaturated adiabats may be replaced by dry adiabats for all practical purposes. An example would show that there would be a difference of about 1°C in the potential temperature if a parcel of air was brought down a dry adiabat instead of the steepest unsaturated adiabat from 400 mb to 1000 mb.

4.13 THE ADIABATIC PROCESSES FOR MOIST SATURATED AIR

In a dry or unsaturated adiabatic expansion, the work done to expand a parcel of air is drawn solely from the internal energy of the gas. However, in a saturated adiabatic expansion, the latent heat of condensation (or sublimation) of water vapour is also available to help expand the parcel, so that a saturated parcel cools less than an unsaturated one, for a given amount of decompression.

If the products of condensation do not escape from a parcel, they are available for re-evaporation in the event of the parcel being compressed. This assumption leads us to the idea of a

■ reversible saturated adiabatic process, defined to be one in which none of the products of condensation escape from the parcel.

At the other extreme we have the

■ pseudo-adiabatic process for saturated air, defined to be one in which all of the products of condensation escape from the parcel.

Since an unknown fraction of the products of condensation escapes from a real parcel of saturated air, the actual adiabatic processes for saturated air lie somewhere between the extremes. However, because the heat capacity of the products of condensation is very small compared with that of the moist air, the two processes described above yield almost identical cooling rates and so the actual adiabatic process for saturated air is represented equally well by the reversible or the pseudo-adiabatic process (at least as far as cooling rates but not necessarily as far as precipitation rates are concerned). The saturated adiabats on the tephigram and the skew $(T, -\log p)$ diagram are therefore constructed using the process which is the simpler from the point of view of computation, and this happens to be the pseudo-adiabatic process. Furthermore, since the condensation of water vapour into water droplets is quite common above the freezing level in the real atmosphere, the pseudo-adiabatic computations are carried out assuming the product of condensation to be water.

4.14 EXACT EQUATION FOR THE RAIN STAGE OF THE PSEUDO-ADIABATIC PROCESS

Let us again take a parcel of saturated air at T, p, x_s. After a small pseudo-adiabatic expansion the air is in the state $(T + dT)$, $(p + dp)$, $(x_s + dx_s)$. Now consider a mass of $1 + x_s$ kilograms of moist air made up of 1 kilogram of dry air and x_s kilograms of water vapour. In the pseudo-adiabatic process the quantity $-dx_s$ of water vapour condenses and drops out as precipitation. The condensation releases the quantity of heat

$$dQ = -L\,dx_s \tag{4.31}$$

which is used to heat the moist air.

It follows from (2.28) that

$$dq = c_p\,dT - RT\frac{dp}{p}$$

This gives the heat absorbed by the moist air per unit mass.

Now

$$dQ = (1 + x_s)\,dq \tag{4.32}$$

Equating the heat released due to the latent heat of condensation of the water vapour to the heat absorbed by the moist air we obtain

$$-L\,dx_s = (1 + x_s)\left(c_p\,dT - RT\frac{dp}{p}\right) \tag{4.33}$$

We may put (4.33) in the form

$$-L\,dx_s = (1 + 1.90x_s)c_{pd}\,dT - (1 + 1.61x_s)R_d T\frac{dp}{p} \tag{4.34}$$

by substituting c_p and R in terms of c_{pd}, R_d and x from (4.27) and (4.25) and multiplying through, ignoring terms in x^2.

4.15 EXACT EQUATION OF THE REVERSIBLE SATURATION ADIABATIC PROCESS

In the reversible saturation adiabatic process the condensed water is retained in the system in the form of cloud droplets.

Let x be the total mass of water substance in a saturated parcel containing unit mass of dry air. The system will then consist of $1 + x_s$ kilograms of moist air and $x - x_s$ kilograms of liquid water.

Let the saturated air be in a state T, p, x_s and let it be expanded to a state $(T + dT)$, $(p + dp)$, $(x_s + dx_s)$.

Then, as for the pseudo-adiabatic process, $dQ_1 = -L\,dx_s$. This is the heat released by condensation which is used to heat the moist air, and also

$$dQ_2 = -c_w(x - x_s)\,dT$$

This is the heat given off by the cooling of $x - x_s$ kilograms of liquid water through dT degrees as the system is cooled by this amount as a result of the expansion. The total heat $dQ = dQ_1 + dQ_2$ is absorbed by the moist air:

$$dQ = -L\,dx_s - c_w(x - x_s)\,dT$$

$$= (1 + x_s)\left(c_p\,dT - RT\frac{dp}{p}\right)$$

from (4.32); equating the two expressions, we have

$$-L\,dx_s - c_w(x - x_s)\,dT = (1 + x_s)\left(c_p\,dT - RT\frac{dp}{p}\right) \qquad (4.35)$$

It is seen that the above expression is the same as (4.33) for the pseudo-adiabatic process apart from the addition of the second term on the left hand side of the equation which is due to the cooling of the liquid water retained in the system during the expansion.

If we convert c_p and R into c_{pd}, R_d and x_s as before, (4.35) can be transformed into the form

$$-L\,dx_s = [1 + 1.90x_s + 4.17(x - x_s)]c_{pd}\,dT - (1 + 1.61x_s)R_dT\frac{dp}{p} \qquad (4.36)$$

It is seen that there is a different reversible saturation adiabat through T, p for each value of x, the total water content. They only differ slightly from one another and from the pseudo-adiabats.

A numerical solution of the above equations shows that the pseudo-adiabatic process cools at a slightly faster rate than the reversible process because of the loss of heat content of the precipitated water. The difference is very small and is negligible compared with the effects of turbulence and radiation. Either equation may be used to calculate the adiabatic process of saturated air, provided the process is an expansion. The difference occurs when expansion is followed by compression. In the pseudo-adiabatic process the compression follows a dry adiabat. In the reversible process the condensed water remains in the air and the compression returns along the path of expansion.

In practice we need not use the exact forms of the equations of the adiabatic processes. They can be replaced by a simpler form which will now be developed.

4.16 SIMPLIFIED EQUATION OF THE ADIABATIC PROCESS OF SATURATED AIR

Let a parcel of $1 + x_s$ kilograms of saturated moist air be in state p, T, x_s and expanded adiabatically to $(p + dp)$, $(T + dT)$, $(x_s + dx_s)$ as before. We will now make the slightly incorrect assumption that the latent heat $-L \, dx_s$ is used exclusively to heat the kilogram of dry air, ignoring the heating of the water vapour.

Then

$$dq = c_{pd} dT - R_d T \frac{dp}{p}$$

from (2.28). Equating the latent heat released by condensation to the heat absorbed by the dry air we have

$$-L \, dx_s = c_{pd} \, dT - R_d T \frac{dp}{p} \tag{4.37}$$

It can be seen that the exact equations derived in (4.34) and (4.36) reduce to (4.37) when the correction factors to R_d and c_{pd} are neglected.

Equation (4.37) can then be used to describe the saturation adiabatic process rather than the exact forms.

We may write (4.37) in the form, from (2.45),

$$-\frac{L \, dx_s}{T} = c_{pd} \frac{dT}{T} - R_d \frac{dp}{p} = c_{pd} \frac{d\theta}{\theta}$$

$$c_{pd} \frac{d\theta}{\theta} = d\phi \tag{4.38}$$

from (2.47). It is noted that a saturated adiabatic expansion involves a change in entropy for the moist air.

4.17 ISOBARIC WARMING AND COOLING

In the preceding sections heating and cooling by adiabatic processes involving expansion and compression have been discussed. It is also necessary to consider heating and cooling at constant pressure. Such a process is called isobaric heating and cooling.

Consider a parcel of air in a state T, p, x. Let vapour be condensed from or water evaporated into the parcel. Let either process take place at constant pressure, the latent heat being supplied to or taken from the air. In the case of evaporation the change in mixing ratio is positive and the air provides the latent heat by cooling. In the case of condensation dx is negative and the air absorbs the latent heat by warming. We assume, as in the preceding section, that the heat is used exclusively to heat the kilogram of dry air. Then

$$dq = c_{pd}(dT)_p$$

from (2.28), or

$$-L\,dx = c_{pd}(dT)_p \tag{4.39}$$

Numerically, the above equation states that at constant pressure adiabatic condensation of one part per thousand of water vapour will warm moist air $2\frac{1}{2}\,°C$. Similarly adiabatic evaporation will cool the air $2\frac{1}{2}\,°C$.

4.18 HYGROMETRIC EQUATION

The principle of isobaric cooling by the evaporation of liquid water, the latent heat being supplied by the air, is the basis of the measurement of humidity by the dry- and wet-bulb thermometer. If the air is cooled isobarically by the evaporation of liquid water, until it becomes saturated, the temperature reached at the saturation point is called the wet-bulb temperature. If this operation is performed on $1 + x$ kilograms of moist air composed of 1 kilogram of dry air and x kilograms of water vapour we have, from (4.39),

$$\int_T^{T_w} c_p(1+x)\,dT = -L\int_x^{x_w} dx$$

where T_w is the wet-bulb temperature and x_w is the saturation mixing ratio at T_w.
   Integrating,

$$(c_{pd} + \bar{x}c_{pv})(T_w - T) = L(x - x_w)$$

where $\bar{x}$ is the mean mixing ratio during the process; $\bar{x}c_{pv}$ may be neglected compared with c_{pd}. Approximately

$$T - T_w = \frac{L(x_w - x)}{c_{pd}}$$

and

$$x = \frac{0\cdot622e}{p}$$

approximately. Then

$$T - T_w = \frac{(e_w - e)(0.622L)}{pc_{pd}} \tag{4.40}$$

$T - T_w$ is the wet-bulb depression which is measured by a wet- and dry-bulb thermometer or by a psychrometer. Since e_w is known for different temperatures the vapour pressure e, and consequently the relative humidity, can be calculated. Hygrometric tables of values of the humidity and the dewpoint have been calculated for different values of the wet-bulb depression. Equation (4.40) applies to the evaporation or condensation from the vapour to the liquid phase. If the temperatures are below 0°C and sublimation from the solid to the vapour phase occurs the value of L will be that for the ice–vapour transformation. In this case e_w would be replaced by e_i, the saturation vapour pressure over ice at the wet-bulb temperature.

4.19 CONSTRUCTION OF SATURATION ADIABATS

Equation (4.39) can be used in constructing saturation adiabats on any thermodynamic diagram. In Fig. 4.6 let saturated air be in state p, T, x_s at a point A. Let dx_s be a fixed convenient quantity. We wish to find the point B where the saturation adiabat through A crosses the saturation mixing ratio line $x_s + dx_s$. We first follow the process AA′, heating the air at constant pressure by the amount caused by the latent heat released by the condensation of $-dx_s$ kilograms of water vapour. This amount may be calculated from (4.39) and the resulting temperature defines a point A′. We then follow the dry adiabat through A′ until the saturation value of the air is reached where the dry adiabat intersects the saturation mixing ratio line $x_s + dx_s$ at point B. Note that in Fig. 4.6 dp, dT and dx_s are all negative. The saturation adiabat may then be constructed through AB, and so on through BC, CD, etc.

It will be useful to define some further temperatures which can be found on thermodynamic diagrams:

■ The equivalent temperature is the temperature reached isobarically when all the vapour in a sample of moist air has been condensed.
■ The equivalent potential temperature is the temperature reached by expanding a parcel of air along the saturated adiabatic line until it is completely dried out and then compressing it along the dry adiabat to the pressure of 1000 hPa.
■ The dewpoint temperature is the temperature at which a parcel of air would become saturated if it were cooled isobarically without any change in the mixing ratio.
■ The wet-bulb temperature is the lowest temperature to which air may be cooled by evaporating water into it.

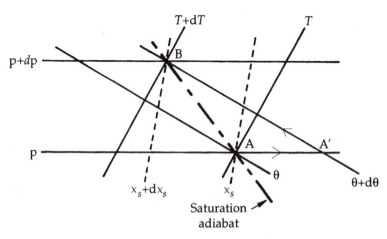

FIGURE 4.6 Graphical construction of saturation adiabats.

■ The wet-bulb potential temperature is the wet-bulb temperature reached when a parcel of air is brought along the saturated adiabat from its wet-bulb temperature to a pressure of 1000 mb.

4.20 NORMAND'S THEOREM

The preceding exercise describing the construction of saturation adiabats illustrates an important principle. It is that the dry adiabatic line, the saturated adiabatic line and the line of equal mixing ratio for a given parcel of air all intersect at a common point, the lifting condensation level. This principle is known as Normand's theorem after Sir Charles Normand who first directed the attention of meteorologists to it.

4.21 SOME USEFUL EMPIRICAL RELATIONSHIPS

Although the hygrometric equation (4.40) is the basic equation for determining humidity, it is not always immediately practical for calculations of dewpoint temperature, relative humidity and mixing ratio. Below are some approximate formulae from Abbott and Tabony (1985) linking the various quantities for use if $T_w > 0°C$. We use T for temperatures (in degrees Celsius), T_D for dewpoint temperature and T_w for wet-bulb temperature; e for vapour pressure and e_s for saturation vapour pressure; and p for pressure in hPa. Note that e is identically $e_s(T_D)$.

The Magnus equation is

$$e_s(T) = 6.107 \exp\left(\frac{17.38\,T}{239.0 + T}\right)$$

The Regnault equation is

$$e = e_s(T_w) - 0.000\,799 p(T - T_w)$$

and the transposed Magnus equation is

$$T_D = \left(\frac{239.0 B}{17.38 - B}\right)$$

where

$$B = \ln(e/6.107)$$

The relative humidity RH is

$$RH = e/e_s(T)$$

and the mixing ratio, x, in $g\,kg^{-1}$, is

$$x = 620\frac{e}{p - e}$$

Given T and T_w, one uses the Magnus equation (twice) to find the saturation vapour pressure at T and at T_w, then the Regnault equation to find $e_s(T_D)$, then the transposed Magnus equation to find T_D, and the last two equations to determine relative humidity and mixing ratio. If $T_w < 0°C$ then the coefficients 22.44 and 272.4 are used instead of 17.38 and 239.0 respectively in the above equations.

4.22 PROBLEMS

1. One sometimes hears the expression 'it makes my blood boil!' At what pressure level in the atmosphere would this actually occur? Normal temperature of the human body is about 37°C.
2. What would be the difference in the latent heat of transformation from water vapour to liquid water if the temperature was 100°C instead of 0°C?
3. Using equation (2.5) derive a more accurate form of Clapeyron's equation than (4.14) and compute the saturation vapour pressure of water at 100°C.
4. From the definitions of mixing ratio and specific humidity show that $x = s/(1 - s)$.
5. Show that $-(L_{wv}/T)\,dx_s = c_{pd}\,d(\log_e \theta)$, which can obviously form the basis of saturated adiabat construction on a tephigram.
6. Construct the saturated adiabat through (1000 mb, 22.5°C) between 1000 mb and 800 mb on an available aerological diagram.
7. Show how Fig. 4.6 illustrates Normand's theorem.
8. Integrate the Clausius–Clapeyron equation for the equilibrium between water and water vapour in the case where the latent heat depends upon temperature, i.e. $L_{w,v} = L_0 - a(T - T_0)$. Show that your answer reduces to (4.21) as $a \to 0$ and hence $L_{w,v} \to L_0$.

5

HYDROSTATIC EQUILIBRIUM

5.1 WHAT IS HYDROSTATIC EQUILIBRIUM?

We have found that for meteorological purposes it is a close enough approximation to treat dry air as if it were a perfect gas. The atmosphere is therefore compressible. The weight of the atmospheric column above some reference height compresses, literally squashes, the column below the reference height so that the mass of the latter column is squeezed together and occupies less space, when averaged over the earth's surface, than the upper layers above that reference point. This is what is meant by hydrostatic equilibrium. It is a stable state in which no vertical motion occurs. Numerical models of some aspects of atmospheric motion are often described as hydrostatic or non-hydrostatic, depending on whether the condition of hydrostatic equilibrium is maintained at all stages of the calculations, or whether that condition is disobeyed and vertical instabilities are allowed to generate. Such vertical instabilities might be heavy showers and thunderstorms. These kinds of weather events usually occur over a relatively small area at any given time. It has been estimated that unstable upward vertical motion only occurs over about 1% of the earth's surface at any given moment. Much of the time there is very gentle subsidence or downward motion of a few centimetres a second. This condition gives rise to quiet, fine weather. Thus hydrostatic equilibrium is a very useful, simple and practical condition to impose.

5.2 THE HYDROSTATIC EQUATION

Thus, the atmospheric pressure measured by a barometer at any point in the atmosphere represents the total weight of an air column of unit cross-section above that point reaching to the outer limits of the earth's atmosphere. We will consider a thin slice of such a column (Fig. 5.1). If we denote the height and pressure at the bottom of the slice by z, p and at the top by $z + dz$, $p + dp$, respectively, the pressure difference dp is the weight of the unit air column of thickness dz.

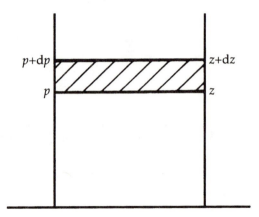

FIGURE 5.1 The hydrostatic equation.

It follows that $dp = -\rho g\,dz$ where dp is the change of pressure along the vertical height axis, ρ is the density and g is the acceleration of gravity; dp is negative if the pressure decreases as the height increases. The z axis extends vertically upwards from the earth's surface and is normal to the plane tangent to the earth's surface at the point of intersection.

The relation $dp = -\rho g\,dz$ is called the hydrostatic equation. It is usually expressed in partial differential form

$$\frac{\partial p}{\partial z} = -\rho g \tag{5.1}$$

The hydrostatic equation represents the balance between the weight of unit mass of air on the one hand and its buoyancy on the other. If (5.1) is valid the atmosphere is said to be in hydrostatic equilibrium.

If the term $-(1/\rho)(\partial p/\partial z)$ is greater than g the air parcel will rise while if it is smaller it will sink. The same principle governs whether an object will float or sink in water or other liquid medium.

5.3 DEFINITION OF LAPSE RATE

The lapse rate is defined as the change of temperature with height. The temperature normally decreases with height in the troposphere. However, sometimes in the boundary layer near the surface, and occasionally at other levels of the atmosphere, the temperature may remain constant or increase with height over short vertical distances. When the temperature increases with height, we define the temperature plot with height an inversion. Inversions often occur in the early morning after a clear cold night during which the air near the ground cools faster than the overlying air. Inversions can be readily designated on an aerological diagram as the slope is upwards and to the right of the isotherms instead of to the left. Mathematically, we denote the lapse rate by the relation $-dT/dz = \gamma$. Therefore, if the temperature decreases with height, which it usually does, we have a positive lapse rate. If there is an inversion the lapse rate would be negative.

5.4 THE THICKNESS EQUATION

We will substitute in (5.1) the value of ρ obtained from the gas equation (2.7). Using the total instead of the partial differential form we have

$$\frac{\mathrm{d}p}{p} = -\frac{g\,\mathrm{d}z}{RT} \tag{5.2}$$

Eq. (5.2) may be integrated between specified limits if it is assumed that the temperature remains constant with height. Then

$$\ln p = -\frac{gz}{RT} + \text{constant}$$

$$z - z_0 = \frac{RT}{g} \ln \frac{p_0}{p} \tag{5.3}$$

where z_0, p_0 are the values of height and pressure, respectively, at the lower or reference level of the layer considered. We see that the term $\ln (p_0/p)$ represents the 'squeezing' of the layer by the pressure p exerted at the upper boundary. In order to obtain a numerical result for the difference $z - z_0$, which is clearly the thickness of the layer, we must keep T constant. This is not strictly possible, unless the layer is isothermal, which would be an unlikely condition. We know from observations that temperature normally decreases with height. However, we may manipulate the situation to our advantage by defining T in equation (5.3) as a mean temperature of the layer. This may be found graphically from an aerological diagram upon which an actual temperature ascent curve has been plotted. We will return to that later. The dependence of the thickness on the mean temperature T_m is just a manifestation of Charles' law, that the volume of a given mass of gas will expand if the temperature is increased and contract if the temperature is decreased. Thus the thickness of a layer is greater if the temperature is higher and smaller if the temperature is lower than some arbitrarily chosen reference value. Equation (5.3) is known as the thickness equation and it is extremely important in the analysis of upper air synoptic charts, a subject we will cover in a later chapter.

5.5 PRESSURE–HEIGHT FORMULAE IN MODEL ATMOSPHERES

By a model atmosphere we mean one which is assumed to have an idealized structure, as far as temperature and humidity are concerned. One of the simplest models we can construct is that in which temperature is constant with respect to height and from which water vapour is absent. One such model is the isothermal model described by equation (5.3). Another form of (5.3) is

$$p = p_0 \exp\left(\frac{-g(z - z_0)}{R_d T}\right) \tag{5.4}$$

T is specified and constant in equation (5.4), not a mean value such as is used in (5.3) to calculate the thickness of a selected layer in the atmosphere. It can be easily seen from (5.4) that it would not be possible to have an isothermal atmosphere of finite thickness, since at the top of the atmosphere $p = 0$. (See problem 1.)

5.5.1 Dry atmosphere with a constant lapse rate

We have already seen that the lapse rate is defined by

$$\gamma = -\frac{\partial T}{\partial z} \tag{5.5}$$

We may then express the temperature at any height z by the relation

$$T = T_0 - \gamma z \tag{5.6}$$

where T_0 is the temperature at the reference level $z = 0$ and γ is a constant lapse rate. We may substitute (5.6) in (5.2). Thus

$$\frac{\mathrm{d}p}{p} = \frac{-g\,\mathrm{d}z}{R(T_0 - \gamma z)}$$

Then the above equation may be separated. We obtain

$$\frac{\mathrm{d}p}{p} = \frac{g}{R\gamma}\frac{\mathrm{d}(T_0 - \gamma z)}{(T_0 - \gamma z)}$$

and integrating,

$$\ln p = \frac{g}{R\gamma}\ln(T_0 - \gamma z) + \text{constant}$$

If we evaluate between limits p_0, p and z_0, z,

$$\ln\frac{p}{p_0} = \frac{g}{R\gamma}\ln\frac{T_0 - \gamma z}{T_0 - \gamma z_0}$$

At the surface $z = z_0 = 0$, $p = p_0$. Then

$$\ln\frac{p}{p_0} = \frac{g}{R\gamma}\ln\frac{T_0 - \gamma z}{T_0}$$

and

$$p = p_0\left(\frac{T_0 - \gamma z}{T_0}\right)^{g/R\gamma} \tag{5.7}$$

Equation (5.7) is the barometric equation for an atmosphere with a constant lapse rate.

5.5.2 Height and lapse rate of a homogeneous atmosphere

One may also integrate the hydrostatic equation, assuming a homogeneous atmosphere, that is an atmosphere in which density does not change with height. In this case

$$\frac{\mathrm{d}p}{\mathrm{d}z} = -\rho_0 g \tag{5.8}$$

where ρ_0 is a constant density. Then

$$\int_{z=z_0}^{z=z} \mathrm{d}z = -\int_{p=p_0}^{p=p} \frac{\mathrm{d}p}{\rho_0 g} \quad \text{and} \quad z - z_0 = \frac{p_0 - p}{\rho_0 g}$$

At the surface $z = 0$, $p = p_0$ and

$$z = \frac{p_0 - p}{\rho_0 g}$$

The height of such an atmosphere may be computed. At the outer limits of the atmosphere $p = 0$ so that

$$z_{TOP} = \frac{p_0}{\rho_0 g} = \frac{RT_0}{g} \tag{5.9}$$

where T_0 is the surface temperature.

If a value of 293 K is substituted in (5.9),

$$z_{TOP} = \frac{287 \times 293}{9.8} \simeq 8.6 \, \text{km}$$

Different values of the height of an atmosphere will be obtained if other values of the surface temperature are assumed. The higher the value of surface temperature substituted in (5.9) the less will be the density and thus the greater the height of the resulting homogeneous atmosphere. Of course, the homogeneous atmosphere never exists in nature; it is purely a hypothetical concept.

If we multiply the hydrostatic equation (5.8) by dT we have

$$\frac{dp}{dz} dT = -\rho_0 g \, dT$$

$$\gamma \, dp = \rho_0 g \, dT$$

Integrating as before since γ is constant

$$\gamma(p - p_0) = \rho_0 g(T - T_0)$$

$$\gamma p - \gamma p_0 = \rho_0 g T - \rho_0 g T_0$$

At the upper boundary $p = 0$ and therefore $T = 0$

$$\gamma = \frac{\rho_0 g T_0}{p_0} = \frac{g}{R_d} = 0.034°\text{C}\,\text{m}^{-1} = 34°\text{C}\,\text{km}^{-1}. \tag{5.10}$$

This is the lapse rate for a homogeneous atmosphere. It will be seen from the following section that the lapse rate for a homogeneous atmosphere of constant density is extremely high and unstable. No such lapse rate could exist in the free troposphere.

Worked Example
What would be the height of an atmosphere of density $1 \, \text{kg}\,\text{m}^{-3}$ at all heights within the air column? Assume surface pressure is 1012 hPa.

Solution:

$$z = 1012 \times \frac{100}{9.8} = 10.326 \, \text{km}$$

5.5.3 The dry adiabatic atmosphere

This is the most useful and practical of all the model atmosphere lapse rates because it is close to the rate at which a parcel, or bubble, which might be a more visual concept, cools as it rises up through the atmosphere. We consider a parcel of dry air which is rising adiabatically in this way. Remembering that the potential temperature remains constant in a dry adiabatic process it follows from (2.45) that

$$\frac{d\theta}{\theta} = \frac{dT}{T} - \kappa_d \frac{dp}{p} = 0$$

or

$$dT = \kappa_d T \frac{dp}{p}$$

Dividing through by dz,

$$\frac{dT}{dz} = \frac{\kappa_d}{p} T \frac{dp}{dz}$$

The dry adiabatic lapse rate will be denoted

$$-\frac{dT}{dz} = \Gamma_d \tag{5.11}$$

Substituting for dp/dz from the hydrostatic equation and using (2.8) and (2.39)

$$\Gamma_d = \frac{g}{c_{pd}} \tag{5.12}$$

If, as before, the lapse rate is considered to be a positive value for a decrease of temperature with height

$$\Gamma_d = 9.8°C \, km^{-1}$$

Now the term dp/dz represents the decrease of pressure of the parcel as it ascends. This must also equal the decrease of pressure of the environment since it is assumed that there is no discontinuity of pressure at the boundary separating the rising parcel or bubble of air from its surroundings.

If the temperature of the environment is not exactly the same as that of the rising air parcel level for level we have

$$\frac{dT}{dz} = \frac{-\kappa_d T \rho' g}{p}$$

where ρ' is the density of the air in the surrounding atmosphere, that is in the environment.

Since there is no discontinuity of pressure $\rho' = p/R_d T'$ and

$$-\frac{dT}{dz} = \gamma_d = \frac{\kappa_d T \rho' g}{\rho' R_d T'} = \frac{g}{c_{pd}} \frac{T}{T'} = \Gamma_d \frac{T}{T'}$$

where T' is now the temperature of the environment and T is the temperature of the rising parcel. Normally $T/T' = 1$ to a close approximation. γ_d is the dry environmental lapse rate for the case when $T \neq T'$.

The height of an atmosphere having a dry adiabatic lapse rate may be calculated from (5.11). Thus

$$\frac{\mathrm{d}T}{\mathrm{d}z} = -\Gamma_\mathrm{d}$$

$$\mathrm{d}T = -\frac{g}{c_\mathrm{pd}}\mathrm{d}z$$

Integrating

$$z - z_0 = \frac{c_\mathrm{pd}}{g}(T_0 - T)$$

At the upper boundary $p = 0$ and therefore $T = 0$, and at the lower boundary $z_0 = 0$. Then

$$z_\mathrm{TOP} = \frac{c_\mathrm{pd}}{g}T_0$$

If $T_0 = 293\,\mathrm{K}$, $z_\mathrm{TOP} = 29.9\,\mathrm{km}$.

5.6 STABILITY AND INSTABILITY

A parcel of air is said to be stable, unstable or indifferent with respect to its environment if, on being given an initial impulse, it returns to its original position, continues its movement or stays where it is.

An example of the concept of stability may be taken from the world of solid things. In the first case we assume that a marble rests at the bottom of a symmetrical concave surface (Fig. 5.2). If the marble is given a flick with the finger it will roll a short distance up the surface, soon returning to its initial position. This is the stable case. In the second case it is assumed that the marble rests on a small perch at the top of a symmetrical convex surface. A flick of the finger will send the marble rolling away down the surface. This is the unstable case. In the third case let it be assumed that the marble rests on a perfectly flat horizontal surface. If the marble is flicked it will move a short distance over the surface and then come to rest and remain where it is. This is the indifferent case.

Now a parcel of air moving through its environment will, if unsaturated, follow the dry adiabatic curve or, if saturated, follow the saturated adiabatic curve. By thus following its path on an upper air diagram its density relative to its surroundings may be seen at a glance. The assumption is made that there is no mixing between the rising parcel and its environment.

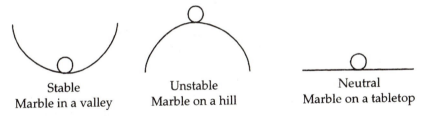

Stable	Unstable	Neutral
Marble in a valley	Marble on a hill	Marble on a tabletop

FIGURE 5.2 Stable, unstable and neutral marble.

From the hydrostatic equation (5.1)

$$\frac{\partial p}{\partial z} = -\rho g$$

or

$$\frac{1}{\rho}\frac{\partial p}{\partial z} + g = 0 \qquad (5.13)$$

showing that the vertical acceleration is zero.

Now if the parcel is not in hydrostatic equilibrium it will have a vertical acceleration

$$\frac{\mathrm{d}^2 z}{\mathrm{d}t^2} = \ddot{z} = -\frac{1}{\rho}\frac{\partial p}{\partial z} - g \qquad (5.14)$$

In (5.14) the term

$$\frac{\partial p}{\partial z} = -\rho' g$$

where ρ' is the density of the surrounding air as distinguished from the density of the rising parcel. If we substitute for $\partial p / \partial z$ in (5.14) it is seen that

$$\ddot{z} = \frac{\rho' g - g\rho}{\rho}$$

or

$$\ddot{z} = g\left(\frac{\rho'}{\rho} - 1\right) = g\frac{(\rho' - \rho)}{\rho} \qquad (5.15)$$

If the air parcel is lighter than the environment $\rho < \rho'$ and $\ddot{z}$ is positive, that is the vertical acceleration is directed upwards since $-(1/\rho)(\partial p/\partial z)$, the buoyancy acceleration, exceeds g. If $\rho > \rho'$, $\ddot{z}$ is negative from (5.15) and is directed downwards. In this case the buoyancy acceleration is less than g, and (5.14) equates the vertical acceleration to the resultant buoyant force acting on unit mass of the parcel.

As density is not directly available from aerological data the buoyant force may be conveniently expressed in terms of the temperature. Substituting from the gas equation in (5.15) and remembering that since $p = p'$ the parcel will adjust its pressure to the pressure of the environment, we obtain

$$\ddot{z} = \frac{g(T - T')}{T'} \qquad (5.16)$$

where T and T' are the temperatures of the parcel and environment respectively; strictly speaking T, T' should be replaced by the corresponding virtual temperatures T^*, $T^{*'}$ as derived in (4.26). In practical meteorology the actual temperature is often used instead of the virtual temperature since the difference between the two is small.

From (5.5) the lapse rate was defined as $\gamma = -(\partial T/\partial z)$. We may call γ' the environment lapse rate and define $\gamma = -(\partial T/\partial z)$ as the individual lapse rate. At the reference level it is assumed that the parcel and its environment have the same thermodynamic properties. When the parcel is given an impulse it will separate from its environment and its thermodynamic properties may then differ from its environment in any new position it may take up.

Now

$$\frac{\mathrm{d}T}{\mathrm{d}z} \cdot z = -\gamma z = T - T_0$$

where T_0 is the temperature of the parcel at the lower reference level, and

$$\frac{\mathrm{d}T'}{\mathrm{d}z} \cdot z = -\gamma'z = T' - T_0$$

where it is assumed that the temperature of the environment at the lower reference level is also T_0, that is it is the same as that of the parcel at the lower reference level.

Subtracting,

$$(\gamma' - \gamma)z = T - T' \tag{5.17}$$

Substituting in (5.16) we have

$$\ddot{z} = \frac{gz(\gamma' - \gamma)}{T'} \tag{5.18}$$

Thus the value of the vertical acceleration of a parcel of air is a function of the difference between the environment and individual lapse rates. We may consider the case where the reference level is at the ground. Then, if the parcel is given an initial upward impulse,

- $\ddot{z}$ is positive and therefore directed further upwards if $\gamma' > \gamma$
- $\ddot{z}$ is zero when $\gamma' = \gamma$
- $\ddot{z}$ is negative and therefore directed downwards back to the reference level when $\gamma' < \gamma$.

If the reference level is at a higher level, that is above the ground, the sign of $\ddot{z}$ will be the same as above for an upward-directed impulse. The sign of $\ddot{z}$ will be reversed for a downward-directed impulse. Thus, if $\gamma' > \gamma$ and the parcel is given an impulse downwards it will continue to accelerate in that direction.

$\gamma' \gtreqless \gamma$ according to whether the lapse rate or decrease of temperature with height in the environment is greater than, equal to or less than the decrease of temperature with height undergone by the individual parcel when it is displaced from the environment.

Now γ represents the process curve lapse rate to which a parcel is subjected during ascent or descent. This is Γ_d, the dry adiabatic lapse rate for dry air, and Γ_s, the moist adiabatic lapse rate for saturated air.

We may now state the following conditions governing stability and instability according to the criteria which were stated at the head of this section. We give a parcel of air an initial impulse upwards; then if $\ddot{z}$, the vertical acceleration, is positive and directed upwards, that is if the parcel continues to rise, then the atmosphere is unstable.

If $\ddot{z}$ is zero the atmosphere is in equilibrium, and if $\ddot{z}$ is directed downwards, that is if the parcel returns to its initial position, the atmosphere is stable. The sign of $\ddot{z}$ would be reversed for an initial downward impulse. Thus, we see from our foregoing conclusions that, for dry air,

(i)   if $\gamma' > \Gamma_d$ the atmosphere as shown by the sounding curve is unstable,
(ii) if $\gamma' = \Gamma_d$ the sounding curve is indifferent, and
(iii) if $\gamma' < \Gamma_d$ the sounding curve is stable;

and, for saturated air,

(iv) if $\gamma' > \Gamma_s$ the sounding curve is unstable,
(v) if $\gamma' = \Gamma_s$ the sounding curve is indifferent or neutral, and
(vi) if $\gamma' < \Gamma_s$ the sounding curve is stable.

If we refer to unsaturated air we say further that,

■ if $\gamma' > \Gamma_d$ the sounding curve is absolutely unstable
■ if $\gamma' = \Gamma_d$ the curve is dry indifferent. It is neutral for dry air and unstable for saturated air.

If $\Gamma_s < \gamma' < \Gamma_d$ the curve is conditionally unstable. It is stable for dry air but unstable for saturated air.

If $\gamma' = \Gamma_s$ the curve is saturated indifferent. It is neutral for saturated air but stable for dry air.

If $\gamma' < \Gamma_s$ the curve is absolutely stable. Figure 5.3 illustrates the stability criteria.

If the reference level is stable the parcel will, as already stated, return towards the equilibrium level after displacement. Its inertia will, however, cause it to pass the reference level so that an oscillation will be set up.

We may define a positive number N such that

$$N^2 = \frac{g(\Gamma_d - \gamma')}{T'} \tag{5.19}$$

Equation (5.18) may then be written as

$$\ddot{z} = -N^2 z$$

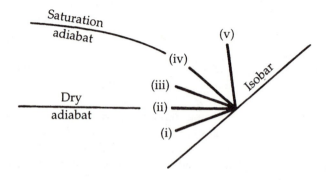

FIGURE 5.3 Stability criteria on the skew $(T, -\log_e p)$ diagram or tephigram.

a solution of which is

$$z = A \sin Nt \tag{5.20}$$

and the parcel will oscillate about the level $z = 0$ with amplitude A and period $2\pi/N$. Such an oscillation is known as a Brent-Väisälä oscillation, after the two persons who first predicted its existence.

As an example, we may compute the period of oscillation of a dry isothermal layer by substituting numerical values for N in (5.19):

$$t = 2\pi \sqrt{\frac{T_0}{g(\Gamma_d - \gamma')}} = 2\pi \sqrt{\frac{T_0 c_{pd}}{g^2}} = \frac{2\pi}{g} \sqrt{c_{pd} T_0}$$

If $T_0 = 273\,\mathrm{K}$, $t = 335$ seconds.

However, since the lapse rate is normally nearer the dry adiabatic the period of oscillation is usually considerably longer.

5.7 ENERGY OF DISPLACEMENT

Equation (5.15) may be stated as

$$\frac{dw}{dt} \cdot dz = \frac{g(\rho' - \rho)}{\rho} dz$$

where $w = dz/dt$ is the vertical velocity. The left hand side is $w\,dw = d\left(\frac{1}{2}w^2\right)$.

This expression represents the change of kinetic energy per unit mass of the particle while it moves through the height dz. When the buoyant force expressed by the term $(g/\rho)(\rho' - \rho)$ is multiplied by dz one obtains

$$\frac{g}{\rho}(\rho' - \rho)\,dz = \frac{g}{\rho}\frac{\rho'}{\rho}(\rho' - \rho)\,dz = -dp\frac{(\rho' - \rho)}{\rho\rho'}$$

$$\frac{dw}{dt} \cdot dz = -dp(\alpha - \alpha') \tag{5.21}$$

The term on the right hand side of (5.21) may be expressed as an area on an (α, p) diagram. On the tephigram the element of area is given by the small parallelograms bounded by the isobars and isotherms since α, α' are proportional to T, T' for a given pressure. The area on a tephigram also represents work, (3.5). Equation (5.21) then represents the element of work done by the buoyant force on unit mass of the parcel while it moves through the height dz.

Integrating (5.21),

$$A = -\int_{p_1}^{p_2} (\alpha - \alpha')\,dp \tag{5.22}$$

This area, A, is that bounded by the two pressure levels p_1 and p_2 and the process and environment curves respectively (Fig. 5.4). If the process curve is warmer than the environment curve the air is unstable and the area described represents the positive latent energy of instability within the layer. This energy is released when a suitable impulse or trigger action sets off the vertical motion. A rising bubble of air may then remain warmer than the air through which it is ascending over a considerable height range. Convective weather phenomena such as showers and thunderstorms arise in this way. If the process curve is colder than the environment curve the atmosphere is stable and the area represents the amount of work which must be performed on unit mass of the parcel to displace it from level p_1 to level p_2 or vice versa. The energy is then negative. The magnitude of the positive or negative latent energy in the atmosphere is a useful means of judging its relative stability. Isopleths of such values may be plotted on synoptic charts and the resultant patterns analysed and used as a further tool in the complex technique of weather forecasting.

In Fig. 5.4 the energy is represented in schematic form by the shaded area. In this example the air is dry. If it is unsaturated it may be assumed to be dry for practical purposes. The process curve is the dry adiabatic line and the environment curve is the plot of the actual temperatures at different pressures or height levels measured by a radiosonde instrument.

On occasions the lapse rate may be stable for dry air and unstable for saturated air. The atmosphere is then called conditionally unstable. If, in such a case, heating of relatively moist air at low levels during a warm day forms a superadiabatic lapse rate the air at these levels may ascend until it becomes saturated and then ascend further along the saturated curve. Forcible lifting over a hill or mountain barrier may also give the required initial impulse instead of solar heating. If the impulse is insufficient the air will sink back to its original level. A conditionally unstable lapse rate is illustrated in Fig. 5.5.

Charles Normand was an internationally known meteorologist who lived in the first part of the twentieth century. He attempted to classify various types of stratification on the basis of the energy principle inherent in the aerological diagram. In Fig. 5.5 suppose that ACBEG is the process curve for a parcel of air moving in an environment whose virtual temperature is given by ABDEF, and let the lower negative area be of magnitude N and the upper positive area be of magnitude P. Then, Normand's classification of stability distinguishes the following three cases:

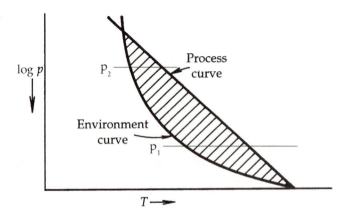

FIGURE 5.4 Area of latent energy.

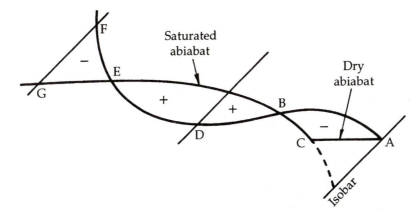

Figure 5.5 Schematic example for conditional instability: environment and adiabatic lapse rate curves.

1. $N > 0$, $P = 0$ absolute stability
2. $N = 0$, $P > 0$ absolute instability
3. $N > 0$, $P > 0$ conditional or latent instability, which can be subdivided into
   $P > N$ real latent instability
   $P < N$ pseudo-latent instability

In the case of absolute stability, there is just no energy available (positive area) to start an initially stationary parcel moving upwards and, in the case of absolute instability, there is no lower negative area to inhibit buoyant vertical motion. The latently unstable cases are more complex, however.

In the real latent case, the amount of energy which must be supplied to a parcel at A in Fig. 5.5 in order for it just to reach B is less than the amount which can be released above B, and the difference $(P - N) > 0$ is taken to be the amount of specific energy available for buoyant convection. In practice, the latter energy appears to be only available on occasions when N is small and the probability of convection occurring increases as $N \to 0$ in this case. One might be tempted to attribute this to the fact that some parcels would be bound to penetrate a very thin lower negative layer as a result of external impulses. The initial velocities required to overcome even a small N are quite large, however, and it is thought to be more likely that, in situations where N is small above a radiosonde station, there may exist nearby superheated areas where it is zero and absolute instability is therefore present. In the pseudo-latent case $(P - N) < 0$ and so there is no net amount of specific energy available for buoyant convection.

In all cases where a parcel arrives at the level E with non-zero kinetic energy, the assumption of the closed-parcel theory is that the parcel will ascend sufficiently far into the upper negative region to lose that kinetic energy and will then execute vertical oscillations about E with the Brunt–Vaisala frequency expressed in equation (5.19). In practice, cloud tops do not usually reach the top of the upper positive area unless they are powerful cumulo-nimbus clouds, when wide anvils stretch out from the tops of the convective part of the clouds. In conclusion of the discussion, it should be borne in mind that there are better and more exact theories of penetrative convection, such as the entrainment theory, which allows for mixing of the convective element with the environment, and the slice theory, which allows for the downward movement of the environment, which must compensate the upward motion of the buoyant elements, if mass is conserved. These more advanced ideas are beyond the scope of the thermodymamics content of this text. They would be incorporated in numerical models. However, for most practical analyses, particularly where the issue of weather forecasts has to meet a time deadline, the parcel methods discussed here will be sufficiently accurate.

5.8 CONVECTIVE AVAILABLE POTENTIAL ENERGY

Figures 5.3–5.5 show examples of different kinds of lapse rates in schematic form. Real examples on an aerological diagram would also show the humidity profile as shown by the mixing ratio lapse rate. In Fig. 5.4 it is assumed that the air is dry, but in Fig. 5.5 it is assumed that the rising parcels of air become saturated at the point C. Below C the parcels rise along the dry adiabat line, but after C, the condensation level, the parcels follow the saturated adiabats. In Fig. 5.5 the area marked with plus signs represents the amount of latent or available energy, as discussed in the preceding section. This energy may also be called convective available potential energy (CAPE). We may define potential energy per unit mass at some level z by the expression

$$g(z - z_0)$$

where z_0 is a reference level. The potential energy of the (z_0, z) layer is then

$$E_p = \int_{z_0}^{z} \rho g(z - z_0)\, dz \tag{5.23}$$

For the whole atmosphere from $z_0 = 0$ to ∞, integration of (5.23) by parts yields

$$E_p = \int_0^{\infty} p\, dz$$

which may be rewritten as

$$E_p = \frac{1}{g} \int_0^{\infty} p\alpha\rho g\, dz = -\frac{1}{g} \int_{p_0}^{0} p\alpha\, dp = -\frac{R}{g} \int_{p_0}^{0} T\, dp.$$

Since the internal energy of such a column is

$$-\frac{C_v}{g} \int_{p_0}^{0} T\, dp,$$

it follows that its total potential energy (TPE) is

$$-\frac{C_p}{g} \int_{p_0}^{0} T\, dp. \tag{5.24}$$

Now suppose that the TPE in our column is somehow minimized by means of a virtual adiabatic re-arrangement of mass. The difference between the actual

TPE and the minimum TPE represents the amount of TPE available for convection and may be regarded and defined as the CAPE. This is a very important property of the atmosphere. In climatic studies and climatic models sea surface temperature anomalies are regarded as very important indicators of rainfall anomalies. Thus, high sea surface temperature anomalies are associated or positively correlated with increases in rainfall and vice versa. It should be borne in mind, however, that it is really CAPE anomalies that are the important criteria. Thus, if sea surface temperatures increase and the upper level temperatures remain constant then the amount of CAPE does increase. Overturning and cloud and precipitation may increase. But if the upper level temperatures change by the same amount as the sea surface temperatures the amount of CAPE will not change much. There has been some debate as to the role of CAPE in climatic change. Professor R. S. Lindzen at MIT has identified CAPE as a parameter which can be used to help in the detection of climate change. He has referred to the climate of 18 000 years ago in the midst of the last glaciation and suggests that on the basis of observations of unicellular creatures called 'forams', the sea surface temperature in the tropics was only about a degree colder than today. However, the temperature at 5 km, estimated from evidence of the height of the snowline on high mountains, was about 5°C colder than today. This meant that there was much more CAPE, so that CAPE changes inversely with sea surface temperature.

5.9 LAPSE RATE FOR UNSATURATED AIR

The adiabatic process for unsaturated air is given by (2.28)

$$c_p \, dT - \alpha \, dp = 0$$

where

$$c_p = c_{pd}(1 + 0.9x)$$

from (4.27). Then

$$c_p \frac{dT}{dz} = \alpha \frac{dp}{dz} = -g$$

$$\Gamma = -\frac{dT}{dz} = \frac{g}{c_p} \tag{5.25}$$

The error arising in neglecting $0.9x$ compared with 1 is small. For all practical purposes we may therefore consider the adiabatic lapse rate for moist unsaturated air to be the same as that for dry air:

$$\Gamma = \Gamma_d = \frac{g}{c_{pd}} \tag{5.26}$$

5.10 LAPSE RATE FOR SATURATED AIR

We may derive the lapse rate which expresses the rate at which moist air will cool as it ascends when it is saturated.

The saturated process is described by (4.33) as

$$-L dx_s = (1 + x_s)[c_p dT - RT d(\ln p)]$$
$$-L dx_s = (1 + x_s)(c_p dT + g dz)$$

Dividing through by dz

$$-L \frac{dx_s}{dz} = (1 + x_s)\left(c_p \frac{dT}{dz} + g\right)$$

$$-L \frac{dx_s}{dz} = (1 + x_s)(-c_p \Gamma_s + g)$$

where Γ_s is the adiabatic lapse rate for saturated air,

$$\Gamma_s c_p - g = \frac{L(dx_s/dz)}{1 + x_s}$$

$$\Gamma_s = \frac{L(dx_s/dz)}{c_p(1 + x_s)} + \frac{g}{c_p} \qquad (5.27)$$

Now $x_s = \epsilon e_s/p$ from (4.47), and, differentiating (4.24),

$$\frac{dx_s}{dz} = \frac{p\epsilon(de_s/dz) - e_s\epsilon(dp/dz)}{p^2} = \frac{-\epsilon \Gamma_s}{p} \frac{de_s}{dT} + \frac{e_s \epsilon g}{pRT}$$

Substituting for de_s/dT from (4.18)

$$\frac{dx_s}{dz} = \frac{-\epsilon L e_s \Gamma_s}{R_v T^2 p} + \frac{e_s \epsilon g}{pRT} = -\frac{x_s L \Gamma_s}{R_v T^2} + \frac{x_s g}{pRT}$$

We assume that $1 + x_s \simeq 1$ and $R \simeq R_d$, $c_p \simeq c_{pd}$. Then, from (5.27),

$$\Gamma_s = \frac{L}{c_{pd}}\left[-\frac{x_s L \Gamma_s}{R_v T^2} + \frac{x_s g}{R_d T}\right] + \frac{g}{c_{pd}}$$

$$\Gamma_s + \frac{x_s L^2 \Gamma_s}{c_{pd} R_v T^2} = \frac{x_s g L}{c_{pd} R_d T} + \frac{g}{c_{pd}}$$

$$\Gamma_s\left[1 + \frac{x_s L^2}{c_{pd} R_v T^2}\right] = \frac{g}{c_{pd}}\left[1 + \frac{x_s L}{R_d T}\right] \qquad (5.28)$$

$$\Gamma_s = \frac{\Gamma_d(1 + x_s L/R_d T)}{1 + x_s L^2/c_{pd} R_v T^2}$$

It is seen that the saturated adiabatic lapse rate is a fraction of the dry adiabatic lapse rate and that this fraction is a function of the amount of water vapour in the air. When the air is quite dry $x_s = 0$ and $\Gamma_s = \Gamma_d$, the dry adiabatic lapse rate. The saturated adiabatic lapse rate at different levels may be evaluated from (5.28) for given initial conditions of p, T. It is seen that at very high levels the saturated adiabats on a tephigram approach the dry adiabats.

5.11 PROBLEMS

1. Show that $p = p_0 \exp(-\text{constant} \cdot z)$ for an isothermal atmosphere. Why is such an atmosphere unlikely to exist?

2. Figures 5.4 and 5.5 are schematic. Construct ascents of actual cases of temperature and humidity profiles and shade the areas of CAPE.

3. At a particular instant, a dry parcel of air has a temperature of 250 K and is moving vertically through a dry isothermal environment of the same temperature. Under the assumption that the motion of the parcel is adiabatic, find its period of oscillation about the initial reference level. Suppose the environment lapse rate is dry adiabatic.

4. If a parcel has an initial upward velocity w'_0, what negative area can it describe before its kinetic energy is consumed? Into what type of energy is this kinetic energy converted?

5. Write a computer program to construct the saturated adiabatic lapse from 40°C to −40°C. Use tables to obtain the values of the saturation mixing ratios at different temperatures.

6. If the lapse rate is constant how does the density change with height? What is the lapse rate when the variation of density with height is zero?

7. Solve equation (5.7) for z, the height. What is the value of the height for the case of a dry adiabatic lapse rate? Express this in terms of the potential temperature, assuming $p_0 = 1000$ hPa.

8. Suppose the atmosphere has a constant lapse rate and surface temperature T_0. Suppose the temperature varies everywhere with time while p_0 is constant. At what height is $\partial p / \partial t$ a maximum?

9. An atmospheric column with a dry adiabatic lapse rate is heated by 1°C throughout. What is the maximum pressure change occurring within the column? Assume $p_0 = 1000$ hPa, $T_0 = 293$ K.

6

THE EQUATIONS OF MOTION: I
THE CORIOLIS FORCE

6.1 INTRODUCTION

So far, the only dynamical phenomenon with which we have dealt is that of convection. In equation (5.14) we introduced the concept of a vertical acceleration for a parcel which was not in hydrostatic equilibrium. We will now examine the idea of force and acceleration in much more detail, concentrating on the way in which horizontal motion of the atmosphere is generated.

The procession of weather changes which takes place from hour to hour, from day to day and from month to month over the surface of our globe is fundamentally the result of the motion of the air, a motion resulting from the action of various forces upon the air parcels. The primary origin of these forces is the energy received from the sun. This energy heats the atmosphere and drives the atmospheric engine. Water vapour is evaporated into the air and subsequently precipitated as dew, frost, rain, hail or snow. The processes involved are complex, but broadly speaking the general circulation of the atmosphere arises as a result of the unequal seasonal and latitudinal and geographic heating of the earth's surface and atmosphere and of the rotation of the earth. The result is that the radiant energy of the sun is transformed into kinetic energy of moving air or wind. The vertical component of the wind is usually small and is often neglected in comparison with the horizontal components. Nevertheless, vertical motion is the prime cause of nearly all forms of cloud and measurable precipitation, and of the absence of clouds, and therefore the main cause of all the weather we experience.

In order to proceed with our arguments we must invoke Newton's second law of motion: the rate of change of momentum of a body is proportional to the impressed force on that body and takes place in the direction of that force. We may write this law in the form

$$\frac{\mathrm{d}}{\mathrm{d}t}(m\mathbf{v}) = \mathbf{F} \qquad (6.1)$$

where m is the mass of the body, $\mathbf{v}$ is its velocity and $\mathbf{F}$ is the resultant impressed force. Note that in this introductory equation we have used $\mathbf{v}$ where the bold type signifies that the velocity is a vector, that is the notation implies that the velocity has both speed and direction. If we assume that the mass of the body stays constant we may write (6.1) in the form

$$\mathbf{F} = m \frac{d\mathbf{v}}{dt} \tag{6.2}$$

As a special case of (6.2) we have Newton's first law of motion: every body continues in a state of rest or of uniform motion in a straight line, except in so far as external impressed forces change that state.

6.2 MOTION AS OBSERVED WITH REFERENCE TO A FIXED FRAME OF COORDINATES

We must first remember that Newton's laws of motion are only valid for a fixed system of coordinates. Such a frame of reference is known as an inertial frame. Whilst we know from relativity that there does not exist an absolutely fixed frame of reference, it turns out that a reference system based upon the 'fixed' stars is sufficiently close to an inertial one for most geophysical purposes. In mathematical physics the most commonly used system of coordinates is the rectangular or Cartesian system of coordinates. In such a system the x, y axes are constructed at right angles to one another on a horizontal plane. The z axis is constructed at right angles and vertically upwards from the origin. In such a system we may write (6.2) in its component forms

$$F_x = m \frac{du}{dt} \tag{6.3a}$$

$$F_y = m \frac{dv}{dt} \tag{6.3b}$$

$$F_z = m \frac{dw}{dt} \tag{6.3c}$$

where F_x, F_y, F_z are the forces acting along the x, y, z axes and u, v, w are the wind components along those axes. In meteorology u and v may be thought of as westerly (i.e. eastwards) and southerly (i.e. northwards) wind components, respectively, while w may be thought of as the vertical (upward) wind speed. x and u are positive if measured to the right of the origin of the rectangular coordinate system and negative if measured to the left of the origin. y and v are positive if they are measured up in the horizontal plane, that is towards the north, and negative if they are measured down in the horizontal plane, that is towards the south. The vertical direction is positive if it is measured up and out of the horizontal plane, and negative if it is measured down into the horizontal plane. It is important to be clear about these basic definitions before proceeding further. All of the equations of motion will contain these symbols. Incidentally, it might be noted that the quantity 'force' is difficult to comprehend, except within

Newtonian mechanics. We all think that we know what it signifies, but it can really only be measured in terms of the acceleration it gives to unit mass. In Chapter 5 we introduced the acceleration of gravity, g, in the hydrostatic equation. g is measured as about $9.8\,\mathrm{m\,s}^{-2}$; that is, it imparts an acceleration of $9.8\,\mathrm{m\,s}^{-2}$ to a body in free fall, without any retarding force such as friction of the air. We cannot really identify force any more closely than this. In future we will drop the symbol m in (6.1)–(6.3). The forces then become forces per unit mass, that is they have the dimensions of acceleration.

6.3 MOTION AS OBSERVED IN A ROTATING FRAME OF COORDINATES

In a fixed system of coordinates such as we have described the resulting motions are fairly simple. However, the surface of the earth is not fixed in space. It describes an orbit around the sun and this gives rise to our seasons of summer and winter. But more important in the present context is the fact that the earth rotates about its axis with an angular velocity Ω equal to about $7.29 \times 10^{-5}\,\mathrm{rad\,s}^{-1}$ which, of course, is once every 24 sidereal hours. Thus, motion cannot be described properly by (6.3), except at the equator; we shall see the reason for this later. We have to adjust our equations by including another term. This new term expresses an apparent force which arises in consequence of the rotation of the frame of reference. In other words, we must write our equations of motion with reference to a rotating reference frame, and not a reference frame which is fixed in space. The new force is called the Coriolis force. It is a factor which must be included whenever the motion of the air is the subject of study, whether in day-to-day weather forecasting or in climate models. It may only be neglected in studies of small-scale phenomena which do not last longer than an hour or two and which are therefore not influenced a great deal by the earth's rotation. The Coriolis force is often difficult to visualize. Because of this we will look at several simple ways of detecting its existence in everyday life. After these examples we will derive the Coriolis force by several mathematical methods. The examples will become more difficult as we proceed, but the results are the same, although in some examples they show more detail.

6.3.1 The bear and the penguin

The very apt mathematical readers will be asked to excuse this first example. We imagine a polar bear at the exact North Pole. The surface is solid ice and the weather is clear and cloudless and it is the 21st of June. The bear has a good watch and at exactly 12 noon starts to walk towards the sun. Let us assume that $2\,\mathrm{km\,h}^{-1}$ is average bear walking speed. The bear continues to walk until the watch does one complete revolution. The bear has now walked in a straight line towards the sun for 12 hours and thinks it must be a long way from its starting point. Imagine its surprise, that is if the bear does not possess any mathematical ability, to find that as the hands of the watch point to 12 midnight 12 hours later, there are bear footprints ahead, and these footprints look surprisingly like those of the walking bear. What are their origin? Actually the bear has completed a full circle, what is called an inertial circle. Although the bear has walked in a straight line towards the sun, which is a fixed star, the earth has rotated underneath. The plane covering the North Pole has rotated in an anticlockwise direction, but the bear has covered its inertial circle in a clockwise direction. To summarize, the bear has walked in a straight line with reference to an observer fixed in space. But to an observer on the earth, in a rotating frame of reference, the bear appears to have walked in a circle.

Suppose we imagine a similar example in the southern hemisphere, at the South Pole. This time a penguin conducts the same exercise. Conditions are identical except now it is the 21st of December. The result is similar except that the penguin will have described a circle in an anticlockwise direction owing to the fact that the earth's surface has rotated in a clockwise direction around the South Pole.

The results of these two examples suggest that there is a force which acts at right angles to the direction of motion. In the northern hemisphere it acts perpendicular to and to the right of the direction of motion and in the southern hemisphere perpendicular to and to the left of the direction of motion. Because of this effect the Coriolis force is sometimes referred to as a deflective force. It does not change the linear speed of the moving body, but changes its direction.

6.3.2 The carousel or merry-go-round

Two children, Jack and Jill, go for a ride on a merry-go-round, which is going round in an anticlockwise direction. They are riding the horses. Jack gets on one side and Jill on the exact opposite side. Jack has a tennis ball. When the carousel is at full speed he throws the ball across to Jill and shouts at her to catch it. Jack watches the ball and sees it career off to the right far out of Jill's grasp. What has happened? The ball has travelled in a straight line but Jack and Jill are rotating at a relatively fast rate. Consequently, the ball appears to veer off to the right. It is an example of the effect of the Coriolis force.

Suppose now the merry-go-round starts to rotate in a clockwise direction and the same experiment is carried out. The tennis ball will now appear to have veered off to the left as watched by Jack. This experiment simulates the effect of the Coriolis force in the southern hemisphere.

6.3.3 A simple practical example of the Coriolis force

Take a sheet of paper and position it on a desk or table with a pin, in such a way as to allow the paper to rotate. With one hand slowly rotate the sheet of paper in an anticlockwise direction. With the other hand take a pen or pencil and move the point towards a fixed object on the desk or opposite wall, keeping the eye fixed on this object, not on the paper. In this example the pencil is moving in a straight line in space, but the paper is rotating. The path drawn by the pen or pencil will be a curve to the right (Fig. 6.1), another example of the Coriolis or deflective force. The same experiment may be conducted with a record turn-table.

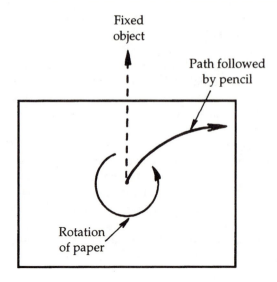

Figure 6.1 Illustration of the deflective force.

6.3.4 Simple mathematical derivation of the Coriolis force

We will again return to the North Pole and imagine that we have there a human observer. This observer can identify a specified object which is shot in a straight line away from the observer with velocity V as illustrated in Fig. 6.2. In the absence of friction or any other retarding force the object would obey Newton's second law and continue moving in a straight line along OA. However, the earth is rotating anticlockwise with angular velocity Ω, so that after an interval of time $\delta(t)$ the observer is facing OA'. The observer would expect to see the object at OA' but, instead, sees it at OA, which is to the observer's right. The object appears to have been deflected from OA' to OA. Now if AA' is small compared

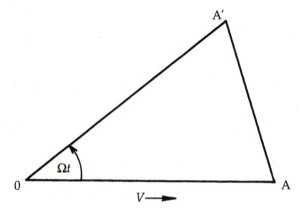

Figure 6.2 Derivation of the expansion for the Coriolis force.

with OA

$$AA' = OA'\Omega = (Vt) \times (\Omega t) = V\Omega t^2$$

$$AA' = \tfrac{1}{2}at^2$$

where a is acceleration, and

$$a = 2\Omega V \tag{6.4}$$

Equation (6.4) expresses the Coriolis force per unit mass acting on a body or parcel of air moving over a plane surface which is rotating about a central axis normal to the surface. We have obtained some additional information from (6.4). It is that the deflective force is proportional to the velocity.

6.3.5 The Foucault pendulum

The action of the Coriolis force can be explicitly illustrated by a Foucault pendulum which is suspended at the end of a very long cable attached to the roof of the ceiling of a vaulted hall. Foucault pendulums are often exhibited in science museums, such as the South Kensington Science Museum in London. Suppose a visitor enters the museum in the morning and observes the pendulum swinging from right to left. Suppose the visitor leaves the museum late in the afternoon and passes the pendulum on the way out. The visitor will notice that now the pendulum is swinging at a different angle, that is in a different vertical plane than it had been earlier in the day. The pendulum has actually not changed its direction of swing in space, but the earth has rotated underneath it so that the pendulum appears to have changed its direction of swing. An example of this is shown in Fig. 6.3.

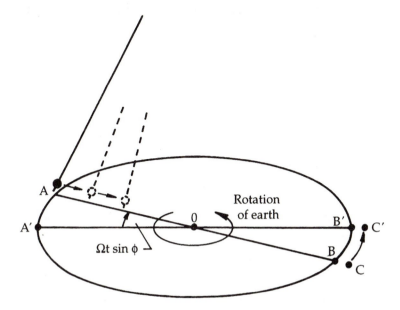

FIGURE 6.3 The Foucault pendulum deflection in the northern hemisphere.

6.4 CONCLUSION

We have given five examples of ways in which the Coriolis force can be detected. Four of these were practical examples and one was an easy mathematical derivation. All of the examples studied were inertial. That is, they assumed that a body or object was moving at some constant speed caused by an initial impulse, or walking at some constant and controlled speed, as in the examples of the bear and penguin. Also we have so far considered that the underlying surface is rotating with some angular velocity Ω and that the axis of rotation is perpendicular to the underlying surface at some selected point which we chose to use as an origin. In the next chapter we will be concerned with air parcels in the atmosphere and will introduce two additional factors. The first of these are forces which act to accelerate (or decelerate) the air parcel along the coordinate axes in accordance with Newton's second law. The second factor is to consider the spherical shape of the earth and break down the magnitude Ω of the earth's rotation to one component around an axis perpendicular to the horizontal plane, that is along the local vertical, and another component perpendicular to the local vertical. The local vertical can be identified as a line which is coincident with a plumb line with a weight on the end of it. We shall then derive the full equations of motion, including the Coriolis terms in a more rigorous manner than hitherto.

6.5 PROBLEMS

1. What is the radius of the circle described by the polar bear?
2. Why does the penguin meet its footprint in 12 hours and not 24 hours?
3. Write an essay describing your own personal understanding of the Coriolis force with examples.

7

THE EQUATIONS OF MOTION: 2 DERIVATION IN VARIOUS COORDINATES

7.1 THE PRESSURE GRADIENT FORCE

The most important force in the equations of motion is the pressure gradient force. This arises as a result of the unequal distribution of the mass of the atmosphere over the surface of the earth. It is represented on weather maps by patterns of isobars which are recognized as pressure patterns such as anti-cyclones and cyclones or depressions. The pressure gradient force is measured in terms of the difference in pressure between two points. An expression may be derived for it in the following way.

Imagine an infinitesimal rectangular box whose sides are parallel to the frame of axes, with a pressure p acting on the face ABCD and a pressure $p + (\partial p/\partial x)\,\mathrm{d}x$ acting on the opposite face EFGH (Fig. 7.1). The sides of the box are $\mathrm{d}x$, $\mathrm{d}y$, $\mathrm{d}z$.

Then the corresponding forces acting on ABCD and EFGH are

$$p\,\mathrm{d}y\,\mathrm{d}z \quad \text{and} \quad \left(p + \frac{\partial p}{\partial x}\mathrm{d}x\right)\mathrm{d}y\,\mathrm{d}z$$

respectively. The resultant force F_x acting on the rectangular box in the x direction is the difference between the two forces.

Thus

$$F_x = p\,\mathrm{d}y\,\mathrm{d}z - p\,\mathrm{d}y\,\mathrm{d}z - \frac{\partial p}{\partial x}\mathrm{d}x\,\mathrm{d}y\,\mathrm{d}z$$

$$F_x = -\frac{\partial p}{\partial x}\mathrm{d}V \qquad\qquad (7.1)$$

$$F_x = -\frac{1}{\rho}\frac{\partial p}{\partial x} \quad \text{for unit mass}$$

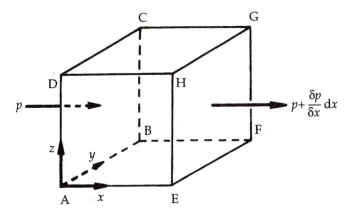

FIGURE 7.1 The pressure gradient force.

This is the pressure gradient force acting in the x direction. If the pressure decreases along the x axis the force is directed along the x axis. If the pressure increases along the x axis the force is directed in the opposite direction, that is towards the origin of the system of coordinates. The forces acting in the directions can be derived similarly. Thus the complete equations for unit mass are

$$F_x = -\frac{1}{\rho}\frac{\partial p}{\partial x}$$

$$F_y = -\frac{1}{\rho}\frac{\partial p}{\partial y} \qquad (7.2)$$

$$F_z = -\frac{1}{\rho}\frac{\partial p}{\partial z}$$

The vertical component F_z is normally balanced by the acceleration of gravity, g. Thus

$$F_z = -\frac{1}{\rho}\frac{\partial p}{\partial z} = g$$

and this is the hydrostatic equation already derived at (5.1). If a parcel of air is not in hydrostatic equilibrium there must be vertical acceleration and (5.1) takes the form of (5.14). The assumption will be made, however, that for the time being vertical motion is negligible in magnitude compared with horizontal motion and accordingly the development of the equations of motion will be confined to horizontal flow.

We may now write

$$\frac{du}{dt} = -\frac{1}{\rho}\frac{\partial p}{\partial x}$$

$$\frac{dv}{dt} = -\frac{1}{\rho}\frac{\partial p}{\partial y} \qquad (7.3)$$

Equations (7.3) refer to horizontal, frictionless flow in a fixed system of coordinates. The resulting motion would be a continuously accelerating velocity directed along the pressure gradient, that is from high to low pressure. Such motion of the atmosphere may occur near the equator as we shall see later, but it is not important since pressure gradients are weak in that region.

7.2 THE SPHERICAL EARTH

In our discussions about the Coriolis force in the previous chapter we contrived the artificial examples where the axis of rotation was perpendicular to the surface. Although this is true at the earth's poles such an assumption is not valid elsewhere. Figure 7.2 is a schematic diagram showing a spherical earth rotating with angular velocity Ω about its axis. It can be readily seen that, if the latitude is denoted by ϕ, then the component of rotation about a point on the local vertical, that is about a line drawn from the point to the centre of the earth, is $\Omega \sin \phi$. It is proportional to the projection of the polar axis of rotation of a line drawn from the centre of the earth O to the specified point on the earth's surface. This projection is shown as OB. Also the component of rotation about a point perpendicular to line OA line is $\Omega \cos \phi$ and this is proportional to the line AB.

We must now substitute Ω in equation (6.4) by $\Omega \sin \phi$. Thus, the Coriolis term becomes $2\Omega V \sin \phi$. The expression $2\Omega \sin \phi$ is called the Coriolis parameter. It will be denoted by f. Having now derived the Coriolis parameter for a spherical rotating earth, and also the pressure gradient terms, we are in a position to write the equations of motion in their basic form.

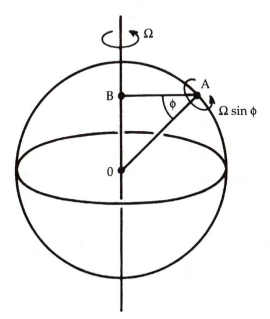

FIGURE 7.2 Component of rotation of a point on the earth's surface about the polar axis.

7.3 THE EQUATIONS OF MOTION

To do this we need to examine Fig. 7.3 to see how the forces act. Remembering that the Coriolis force acts at 90° to the right of the direction of motion of the air parcel in the northern hemisphere, and is proportional to the velocity we may write

$$\frac{du}{dt} = -\frac{1}{\rho}\frac{\partial p}{\partial x} + fv$$

$$\frac{dv}{dt} = -\frac{1}{\rho}\frac{\partial p}{\partial y} - fu$$

The terms on the left hand side of the above equations are the actual residual accelerations along the axes x and y. The equations are normally written in the form

$$\frac{du}{dt} - fv = -\frac{1}{\rho}\frac{\partial p}{\partial x}$$

$$\frac{dv}{dt} + fu = -\frac{1}{\rho}\frac{\partial p}{\partial y}$$

(7.4)

Although we shall derive more rigorous forms of (7.4) the additional terms are neglected for most calculations and simulations of the motion of the atmosphere. Nevertheless, it is necessary to derive the more rigorous forms so that we can decide whether or not the extra terms really can be neglected.

We recall that in the southern hemisphere the terrestrial tangent plane rotates in a clockwise sense when viewed from above. By convention anticlockwise rotation is positive and clockwise rotation is negative. Thus, $f = 2\Omega \sin\phi$ is

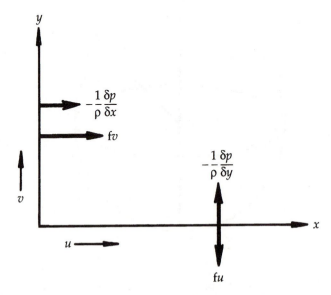

FIGURE 7.3 Equations of motion for horizontal frictionless flow.

positive in the northern hemisphere and negative in the southern hemisphere. This means that in the southern hemisphere equation (7.4) would become

$$\frac{du}{dt} + |f|v = -\frac{1}{\rho}\frac{\partial p}{\partial x}$$

$$\frac{dv}{dt} - |f|u = -\frac{1}{\rho}\frac{\partial p}{\partial y}$$

(7.5)

In describing atmospheric motion in the southern hemisphere as, for example, in numerical general circulation models, the problem of the change in sign is accomplished by setting the latitude angle ϕ negative, so that (7.4) are valid in both hemispheres.

7.4 DERIVATION OF THE COMPONENTS OF THE CORIOLIS FORCE FROM THE LAW OF THE CONSERVATION OF ANGULAR MOMENTUM

We will use one more method to derive the Coriolis parameter. In this case we will derive both the components, that is the component about the earth's axis of rotation and the component about the line perpendicular to that axis; in other words, we shall consider the earth to be a sphere.

First, let us consider that a parcel of air of unit mass is moving with velocity u along a parallel of latitude and let us discuss the relation

$$\left(\Omega + \frac{u}{R}\right)^2 R = \Omega^2 R + 2u\Omega + \frac{u^2}{R}$$

(7.6)

where $R = a\cos\phi$ and a is the radius of the earth. Now the term on the left hand side of (7.6) is the total centrifugal force on the air parcel, and the first term on the right hand side is the centrifugal force due to the rotation of the earth. This is included in the measured value of gravity. The third term is the centrifugal force on the air parcel due to its rotation with velocity u about the axis of the earth. This is small compared with the first term and may be neglected. The term we are interested in is the second term which is the deflective or Coriolis force. This is $2\Omega u$. Now we have already seen that there are two components of Ω: $\Omega\sin\phi$ around the axis of the local vertical, and $\Omega\cos\phi$ around the horizontal vector line connecting the parcel with the axis of the earth. The latter line is perpendicular to the vertical axis. Thus we have

$$\frac{dv}{dt} = -2\Omega u \sin\phi$$

(7.7)

This component of the Coriolis force acts at right angles and to the right of the westerly wind u. It is negative since it is directed towards the south. This is the component we have already met which acts within the horizontal x, y plane. But we also have the other component

$$\frac{dw}{dt} = 2\Omega u \cos\phi$$

(7.8)

which gives rise to a vertical acceleration within the x, z plane. Equations (7.7) and (7.8) both act at right angles to the wind blowing along a parallel of latitude. We will now consider what happens if we displace a parcel northwards (or southwards), that is on the meridional component of the wind, v. To do this we invoke the conservation of angular momentum equation

$$\Omega R^2 = \left(\Omega + \frac{\delta u}{(R + \delta R)} \right)(R + \delta R)^2 \tag{7.9}$$

Expanding (7.9) we may neglect the products of all differentials. We also neglect δR in comparison with R. We find that

$$\delta u = -2\Omega \delta R$$

We have already defined $R = a \cos \phi$ so that $\delta R = -a \sin \delta \phi$.

Therefore $du = +2\Omega a \sin \phi \, d\phi$ and, since $dy = a \, d\phi$, we have

$$\frac{du}{dt} = -2\Omega v \sin \phi \tag{7.10}$$

In a similar manner we can show that if a parcel is projected vertically upward

$$\frac{du}{dt} = -2\Omega w \cos \phi \tag{7.11}$$

Combining equations (7.7)–(7.10) we have the complete set of Coriolis accelerations:

$$\frac{du}{dt} = fv - 2\Omega w \cos \phi$$

$$\frac{dv}{dt} = -fu$$

$$\frac{dw}{dt} = 2\Omega u \cos \phi$$

7.5 DERIVATION OF THE EQUATIONS OF MOTION IN PLANE COORDINATES FROM ROTATING AXES

A more rigorous form of (7.4) may be derived by considering a system of rotating axes referred to axes fixed in space.

In Fig. 7.4 x, y is a system of axes fixed in space with origin O at the North Pole. x', y' are axes fixed to the surface of the earth also with origin at the North Pole, but they rotate with the earth in an anticlockwise direction. Then

$$OM = x \cos \Omega t$$

$$MR = y \sin \Omega t$$

$$PQ = y \cos \Omega t$$

$$RQ = x \sin \Omega t$$

We may describe the rotating coordinates x', y' of a point P in terms of fixed coordinates x, y as follows:

$$x' = x \cos \Omega t + y \sin \Omega t$$
$$y' = y \cos \Omega t - x \sin \Omega t$$

(7.12)

Differentiating with respect to time and using the notation $\dot{x} = dx/dt$, $\dot{y} = dy/dt$, etc., we have

$$\dot{x}' = \dot{x} \cos \Omega t + \dot{y} \sin \Omega t - \Omega x \sin \Omega t + \Omega y \cos \Omega t$$

or

$$\dot{x}' = \dot{x} \cos \Omega t + \dot{y} \sin \Omega t + \Omega y'$$

and

$$\dot{y}' = \dot{y} \cos \Omega t - \dot{x} \sin \Omega t - \Omega y \sin \Omega t - \Omega x \cos \Omega t$$

or

$$\dot{y}' = \dot{y} \cos \Omega t - \dot{x} \sin \Omega t - \Omega x'$$

Differentiating $\dot{x}'$ and $\dot{y}'$ again with respect to time we have

$$\ddot{x}' = \ddot{x} \cos \Omega t + \ddot{y} \sin \Omega t + \Omega \dot{y}' - \Omega \dot{x} \sin \Omega t + \Omega \dot{y} \cos \Omega t$$
$$= \ddot{x} \cos \Omega t + \ddot{y} \sin \Omega t + \Omega \dot{y}' + \Omega^2 x' + \Omega \dot{y}'$$
$$= \ddot{x} \cos \Omega t + \ddot{y} \sin \Omega t + 2\Omega \dot{y}' + \Omega^2 x'$$

(7.13)

and

$$\ddot{y}' = \ddot{y} \cos \Omega t - \ddot{x} \sin \Omega t - \Omega \dot{x}' - \Omega \dot{y} \sin \Omega t - \Omega \dot{x} \cos \Omega t$$
$$= \ddot{y} \cos \Omega t - \ddot{x} \sin \Omega t - \Omega \dot{x}' - \Omega \dot{x}' + \Omega^2 y'$$
$$= \ddot{y} \cos \Omega t - \ddot{x} \sin \Omega t - 2\Omega \dot{x}' + \Omega^2 y'$$

(7.14)

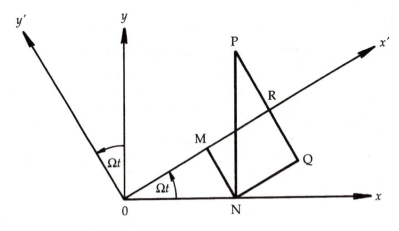

FIGURE 7.4 Rotating frame of reference.

Now from Fig. 7.4 the forces acting at any point along the rotating axes may be expressed in terms of the forces acting along the fixed axes:

$$F'_x = F_x \cos \Omega t + F_y \sin \Omega t$$
$$F'_y = F_y \cos \Omega t - F_x \sin \Omega t$$

(7.15)

where $F_x = \ddot{x}$ and $F_y = \ddot{y}$, for unit mass.

Then

$$F'_x = \ddot{x} \cos \Omega t + \ddot{y} \sin \Omega t$$
$$F'_y = \ddot{y} \cos \Omega t - \ddot{x} \sin \Omega t$$

From (7.13), (7.14) and (7.15) we obtain

$$\ddot{x}' = F'_x + 2\Omega \dot{y}' + \Omega^2 x'$$
$$\ddot{y}' = F'_y - 2\Omega \dot{x}' + \Omega^2 y'$$

(7.16)

Now replacing x', y' by x, y throughout we may write the equations of motion for horizontal and frictionless flow referred to rotating axes fixed to the earth's surface with their origin at the North Pole. For any point located elsewhere on the earth's surface than at the North Pole we must replace Ω by $\Omega \sin \phi$ in the Coriolis terms as already discussed in Section 7.3. Then, from (7.16),

$$\frac{du}{dt} = -\frac{1}{\rho} \frac{\partial p}{\partial x} + fv + \Omega^2 x \sin^2 \phi$$
$$\frac{dv}{dt} = -\frac{1}{\rho} \frac{\partial p}{\partial y} - fu + \Omega^2 y \sin^2 \phi$$

(7.17)

The last two terms are the horizontal component of the centrifugal force due to the earth's rotation. When combined the two terms can be expressed as $\Omega^2 R$. It acts outwards from the axis of rotation, that is, along the line BA in Fig. 7.2. This component is small and absorbed in the measured value of g and is therefore not included separately. The vectorial combination of the gravitational attraction and this small centrifugal force is known as apparent gravity.

7.6 DERIVATION OF THE EQUATIONS OF MOTION IN ROTATING POLAR COORDINATES

So far we have used a plane rectangular coordinate system to express the accelerations in the equations of motion. It is sometimes useful to use polar coordinates as, for example, if we are concerned with circular systems of isobars such as occur in high-pressure systems (anticyclones), or low-pressure systems (depressions, or tropical cyclones or typhoons). For tropical cyclones or typhoons the polar coordinate system is converted to a cylindrical system by including the vertical coordinate z. We will start by transforming the equations (7.3) using the relations

$$x = r \cos \theta$$
$$y = r \sin \theta$$

(7.18)

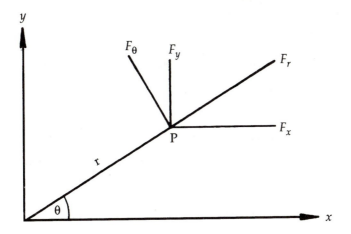

FIGURE 7.5 Transformation from Cartesian to polar coordinates.

where r is the radius vector and θ is the angle r makes with the x axis (Fig. 7.5). Differentiating (7.18),

$$\dot{x} = -r \sin \theta \dot{\theta} + \dot{r} \cos \theta$$

$$\ddot{x} = -r \sin \theta \ddot{\theta} - \dot{\theta}(r \cos \theta \dot{\theta} + \dot{r} \sin \theta) + \ddot{r} \cos \theta - \dot{r} \dot{\theta} \sin \theta$$

$$= \ddot{r} \cos \theta - 2\dot{r} \dot{\theta} \sin \theta - r\ddot{\theta} \sin \theta - r\dot{\theta}^2 \cos \theta \qquad (7.19)$$

and

$$\dot{y} = \dot{r} \sin \theta + r \cos \theta \dot{\theta}$$

$$\ddot{y} = \dot{r} \cos \theta \dot{\theta} + \ddot{r} \sin \theta + r \cos \theta \ddot{\theta} + \dot{\theta}(\dot{r} \cos \theta - r \sin \theta \dot{\theta})$$

$$= \ddot{r} \sin \theta + 2\dot{r} \dot{\theta} \cos \theta + r\ddot{\theta} \cos \theta - r\dot{\theta}^2 \sin \theta \qquad (7.20)$$

Now any force F may be split up into components F_r, F_θ along and perpendicular to the radius vector in just the same way as into F_x, F_y along the x and y axes. From Fig. 7.5 it is seen that we may express F_r, F_θ in terms of F_x, F_y as below:

$$F_r = F_x \cos \theta + F_y \sin \theta$$
$$F_\theta = -F_x \sin \theta + F_y \cos \theta \qquad (7.21)$$

but

$$F_x = \ddot{x} \quad \text{and} \quad F_y = \ddot{y}$$

Therefore

$$F_r = \ddot{x} \cos \theta + \ddot{y} \sin \theta$$
$$F_\theta = -\ddot{x} \sin \theta + \ddot{y} \cos \theta$$

and from (7.19), (7.20) and (7.21)

$$F_r = \ddot{r}\cos^2\theta - 2\dot{r}\dot{\theta}\sin\theta\cos\theta - r\ddot{\theta}\sin\theta\cos\theta - r\dot{\theta}^2\cos^2\theta$$
$$\quad + \ddot{r}\sin^2\theta + 2\dot{r}\dot{\theta}\sin\theta\cos\theta + r\ddot{\theta}\sin\theta\cos\theta - r\dot{\theta}^2\sin^2\theta$$
$$F_r = \ddot{r} - r\dot{\theta}^2$$

and

$$F_\theta = -\ddot{r}\cos\theta\sin\theta + 2\dot{r}\dot{\theta}\sin^2\theta + r\ddot{\theta}\sin^2\theta + r\dot{\theta}^2\sin\theta\cos\theta$$
$$\quad + \ddot{r}\sin\theta\cos\theta + 2\dot{r}\dot{\theta}\cos^2\theta + r\ddot{\theta}\cos^2\theta - r\dot{\theta}^2\sin\theta\cos\theta$$
$$F_\theta = r\ddot{\theta} + 2\dot{r}\dot{\theta}$$

The forces along and tangential to the radius vector are thus, respectively

$$F_r = \ddot{r} - r\dot{\theta}^2$$
$$F_\theta = r\ddot{\theta} + 2\dot{r}\dot{\theta} \tag{7.22}$$

The above expressions refer to fixed polar coordinates. If the coordinate system is rotating as it does when fixed relative to the surface of the rotating earth these expressions must be adjusted. This may be done by letting $\dot{\theta}$ in (7.22) equal $\dot{\theta} + \Omega\sin\phi$. The new $\dot{\theta}$ thus represents the angular velocity relative to axes fixed relative to the surface of the earth:

$$F_r = \ddot{r} - r(\dot{\theta} + \Omega\sin\phi)^2 = -\frac{1}{\rho}\frac{\partial p}{\partial r}$$
$$F_\theta = r\ddot{\theta} + 2\dot{r}(\dot{\theta} + \Omega\sin\phi) = -\frac{1}{\rho r}\frac{\partial p}{\partial\theta} \tag{7.23}$$

Omitting the term for the centrifugal force of the earth as before, we have

$$F_r = \ddot{r} - r\dot{\theta}^2 - fr\dot{\theta} = -\frac{1}{\rho}\frac{\partial p}{\partial r}$$
$$F_\theta = r\ddot{\theta} + 2\dot{r}\dot{\theta} + f\dot{r} = -\frac{1}{\rho r}\frac{\partial p}{\partial\theta} \tag{7.24}$$

which are the required equations.

7.7 DERIVATION OF THE THREE-DIMENSIONAL EQUATIONS OF MOTION IN A SPHERICAL COORDINATE SYSTEM

In order to do this we set up the relations

$$x = r\cos\phi\cos\lambda$$
$$y = r\cos\phi\sin\lambda \tag{7.25}$$
$$z = r\sin\phi$$

as shown in Fig. 7.6, and differentiate twice. The second derivatives of x, y, z are F_x, F_y, F_z, the forces acting along the axes fixed within the earth.

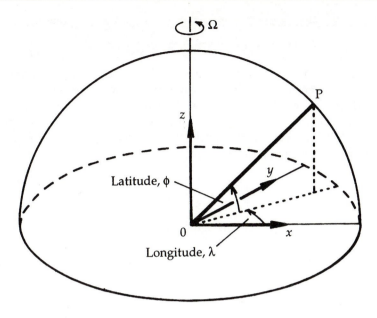

FIGURE 7.6 Illustration of spherical coordinates: λ represents longitude measured anticlockwise around the equatorial plane; ϕ is latitude, the angle between OP and the equatorial plane.

Having found the second derivatives of x, y, z, we deduce from the figure that we must equate the forces along the parallels of latitude and longitude and along the radius vector as follows:

$$F_\phi = -F_x \cos \lambda \sin \phi - F_y \sin \lambda \sin \phi + F_z \cos \phi$$

$$F_\lambda = -F_x \sin \lambda + F_y \cos \lambda \tag{7.26}$$

$$F_r = F_x \cos \lambda \cos \phi + F_y \sin \lambda \sin \phi + F_z \sin \phi$$

The understanding of the relations in (7.26) requires a sense of visual interpretation of the spherical geometry, but the task is made easier by first selecting the simplified cases of $\phi = 90°$ and $\lambda = 0°$ and $\lambda = 90°$. We will leave the actual differentiating and multiplying out as an exercise for the reader. Remembering that the λ in our equations is for a fixed system of coordinates and that for the rotating earth $d\lambda/dt \rightarrow d\lambda/dt + \Omega$, we thus obtain the final form for the equations of motion in spherical coordinates.

We neglect the terms containing $\Omega^2 r$ which represent the centrifugal force of the earth's rotation, which is incorporated in the observed measurement of gravity. Then the final form of the equations of motion in spherical coordinates is

$$r \cos \phi \ddot{\lambda} + 2(\dot{\lambda} + \Omega)(\dot{r} \cos \phi - r \sin \phi \dot{\phi}) = F_\lambda$$

$$r\ddot{\phi} + 2\dot{r}\dot{\phi} + r \cos \phi \sin \phi \dot{\lambda}(\dot{\lambda} + 2\Omega) = F_\phi \tag{7.27}$$

$$\ddot{r} - r\dot{\phi}^2 - r \cos^2 \phi \dot{\lambda}(\dot{\lambda} + 2\Omega) = F_r$$

7.8 EQUATIONS OF MOTION IN TANGENTIAL CURVILINEAR COORDINATES

The equations of motion in spherical coordinates as derived in equations (7.27) are not very convenient to use. We make our observations of wind with reference to the surface of the earth so that a form of coordinate system fixed to the surface of the earth is better. We may then adjust our simple rectangular coordinate system so that it is everywhere tangent to the curved spherical surface of the earth. To do this we set up the relations

$$\mathrm{d}x = r\cos\phi\,\mathrm{d}\lambda$$

$$\mathrm{d}y = r\,\mathrm{d}\phi \tag{7.28}$$

$$\mathrm{d}z = \mathrm{d}r$$

Then

$$u = r\cos\phi\dot{\lambda} \tag{7.29}$$

$$\dot{u} = r\cos\phi\ddot{\lambda} \tag{7.30}$$

$$\dot{\lambda} = \frac{u}{r\cos\phi} \tag{7.31}$$

$$\ddot{\lambda} = \frac{\ddot{x}r\cos\phi - \dot{x}\dot{r}\cos\phi + \dot{x}\dot{\phi}r\sin\phi}{r^2\cos^2\phi} \tag{7.32}$$

Also

$$\dot{\phi} = \frac{v}{r} \quad \text{and} \quad \ddot{\phi} = \frac{r\dot{v} - v\dot{r}}{r^2} \tag{7.33}$$

and $r = a + z$, $\dot{r} = \dot{z} = w$, the vertical velocity.

Multiplying equation (7.32) above by $r\cos\phi$ we have

$$r\cos\phi\ddot{\lambda} = \ddot{x} - \frac{\dot{x}(\dot{z}\cos\phi - v\sin\phi)}{r\cos\phi} \tag{7.34}$$

Substituting the right hand side of this equation into the first relation of (7.27), neglecting z, the height above the earth's surface, in comparison with the radius of the earth, a, and converting the spherical coordinate terms into curvilinear coordinate terms using the relations described, we get

$$\dot{u} + \frac{uw}{a} - \frac{uv\tan\phi}{a} + 2\Omega w\cos\phi - fv = -\frac{1}{\rho}\frac{\partial p}{\partial x} \tag{7.35}$$

This is the first equation in curvilinear coordinates for the accelerations directed along parallels of latitude. Continuing,

$$\dot{\phi} = \frac{v}{r} \tag{7.36}$$

$$r\ddot{\phi} = \dot{v} - \frac{vw}{r} \tag{7.37}$$

and

$$2\dot{r}\dot{\phi} = \frac{2wv}{r} \tag{7.38}$$

Substituting (7.37) and (7.38) into the second relation of (7.27) and making the appropriate conversion to curvilinear coordinates again we have

$$\dot{v} + \frac{vw}{a} + \frac{v^2}{a}\tan\phi + fu = -\frac{1}{\rho}\frac{\partial p}{\partial y} \tag{7.39}$$

Finally, the direct transformation of the third relation of (7.27) is

$$\ddot{z} - \frac{v^2}{a} - \frac{u^2}{a} - 2\Omega u\cos\phi = -\frac{1}{\rho}\frac{\partial p}{\partial z} - g \tag{7.40}$$

We find that we can neglect all terms with the radius of the earth in the denominator as a is large compared with the velocities in the numerator. If we neglect these small terms we end up once again with equations (7.4). These are the equations used in most studies of the motion of the atmosphere.

There is one important thing to remember. All the latter equations we have derived apply to an atmosphere which is inviscid, that is there is no friction. This is an approximation that can be made for motion in the free atmosphere, that is to say, above the boundary layer, which is affected by the friction of the surface, normally greater over land than over the ocean. We will at a later stage discuss the addition of extra terms to include friction. However, for the time being we will only use the equations which describe frictionless motion.

7.9 PROBLEMS

1. Two pedestrians set out together to walk towards a church steeple 10 km away. When they start out a fixed star is seen directly behind the steeple. The first pedestrian walks continuously towards the steeple. The second walks continuously towards the fixed star. If they both walk at a speed of 5 km h^{-1} how far apart will they be when the first has reached the church (a) if the church is at the North Pole? (b) if it is at 45° south latitude? (c) if it is on the equator?

2. Two billiard balls are placed on a billiard table, one at each end at a distance of 10 m apart. The balls are 2 cm in radius. A player strikes one ball directly towards the other. At what speed must the ball travel in order to just miss the ball at the other end of the table? The table is at 43° latitude. Neglect friction.

3. An object is propelled upwards at some starting velocity w_0 at some latitude ϕ. Neglecting friction, where will the object hit the ground on its descent? Find the numerical result for initial upward velocities of 10 m s^{-1} and 50 m s^{-1}.

4. Consider the Coriolis acceleration acting on an air parcel in the plane tangent to the earth's surface and find the velocity and position of the parcel subjected to some initial velocity u_0, v_0. No other forces are present.

5. Derive the equations of motion in spherical coordinates as described in Section 7.7. [Hint: To simplify the double differentiation of the product of three variables set the independent variables in the first and second relations in (7.25) to a, b, c. Differentiate x, y twice using the dot notation. Then differentiate the individual values of a, b, c separately and substitute in your equation for $\ddot{x}$ and $\ddot{y}$ which is in terms of a, b, c.]

8

BALANCED FLOW

8.1 INTRODUCTION

Atmospheric motion is often described as balanced or unbalanced. What do we mean by this terminology? Of course, in the strict sense of the word all motion is balanced. In the equations of motion we have accelerations du/dt, dv/dt and dw/dt. These are essentially residual accelerations along the coordinate axes. When we say the flow is balanced what we really mean is that du/dt, dv/dt and dw/dt are all equal to zero. We will start with the simplest example, which is also the case most frequently used in practical studies.

8.2 THE GEOSTROPHIC EQUATION

We will take the case of straight isobars. If one had not studied the effect of the rotation of the earth one might expect the air to blow across the isobars from high to low pressure. Certainly, if we allow a marble to roll down a smooth slope it appears to run down the hill, not at right angles to the slope. (See problem 1.) However, observations of the wind on a weather map invariably show that the wind appears to blow along the isobars. Why is this? Let us return to equations (7.4). If we set the residual accelerations along the axes equal to zero we have

$$u_g = -\frac{1}{\rho f}\frac{\partial p}{\partial y}$$
$$v_g = +\frac{1}{\rho f}\frac{\partial p}{\partial x}$$

(8.1)

Equations (8.1) are known as the geostrophic equations. If we calculate a value of the density of the air and know the difference of pressure between two selected points on a grid along the x and y axes we can compute the components of the wind speed in $m\,s^{-1}$. In calculating geostrophic winds for weather forecasting purposes, and in some synoptic and climatological diagnostic studies, we may not be interested in the components of the wind, only in the total geostrophic speed along the isobars, irrespective of the direction in which they are oriented. In such a case equations (8.1) reduce to

$$V_g = \frac{1}{\rho f} \frac{\delta p}{\delta n} \tag{8.2}$$

where n is the perpendicular or normal distance across the isobars at the place where that distance is measured. We may always reduce (8.1) to (8.2) by orientating our x axis along the isobars. However, in most diagnostic studies it is better to retain the notation of u and v to represent the west and south winds respectively.

Worked Example
What is the geostrophic wind at 43°N latitude if the pressure gradient is 1 hPa per degree of latitude? Surface pressure is 1012 hPa and temperature 20°C. (One degree of latitude = 60 nautical miles = 111 km.)

Solution:

$$V_g = \frac{1}{\rho f} \frac{\delta p}{\delta n}$$

$$V_g = \frac{100}{111 \times 10^3 \times 1.2 \times 10^{-4}} = 7.5 \, \mathrm{m\,s}^{-1}$$

The relation expressed by (8.2) is sometimes known as the geostrophic assumption. It represents the condition where there is an exact balance between the Coriolis and pressure gradient forces. Equation (8.2) may also be derived very simply by setting the Coriolis force and the pressure gradient force equal to each other as shown schematically in Fig. 8.1(a) for the northern hemisphere and Fig. 8.1(b) for the southern hemisphere. It is an assumption that is confirmed by observations most of the time. If the isobars are curved or if the pressure gradient is changing rather rapidly with time corrections must be made. We will return to this later.

The geostrophic assumption or approximation is extremely useful to meteorologists as it enables an estimate of the wind to be made from a weather map, even though there are no actual wind observations available. This is important because the pressure field is continuous and can be expressed by constructing isobars, whereas the wind cannot be mapped so easily as a continuous field.

It will be noted from equations (8.1) that the geostrophic wind speed increases with decreasing f, that is, with decreasing latitude. At the equator the geostrophic wind speed becomes infinite. This, of course, is absurd. The physical significance of such an impossible mathematical statement is that the assumption that the residual accelerations du/dt and dv/dt are zero breaks down as the equator is approached. The geostrophic assumption is generally not valid in the tropics. However, it may be used satisfactorily for latitudes greater than about 15°N and S. Strictly speaking the geostrophic assumption only applies to east–west and not to north–south isobars. In the former case f is constant, but in the

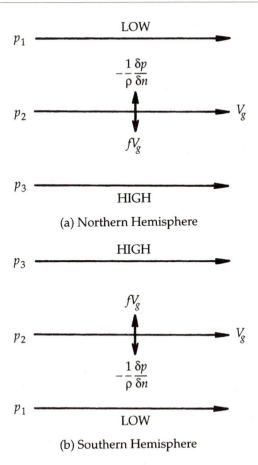

FIGURE 8.1 Balance of forces for geostrophic flow: (a) northern and (b) southern hemisphere. Note in (b) that the absolute value of f is used.

latter case f changes in direct proportion with the latitude so that exact balance cannot be achieved. However, for all practical purposes, particularly in regions removed from the tropics and on scales less than planetary, we can neglect the variation of f with latitude in computing the value of v_g.

Weather forecasters sometimes use a wind scale to calculate the geostrophic wind. The scale consists of a piece of Perspex upon which curved lines are spaced at different distances apart, in order to take into account the latitude. The scale is placed over the isobars on the weather map so that two lines on the scale coincide with two isobars on the synoptic chart at the actual latitude where the wind is being measured. The wind speed may then be read off the scale.

8.3 THE GRADIENT WIND EQUATION

If the isobars are not straight, but curved, a third force must be introduced in addition to the Coriolis and pressure gradient forces. This additional force is the centrifugal force due to the motion of the air around a curved horizontal path, not to be confused with the centrifugal force caused by the rotation of the earth, which is absorbed into gravity. When all three forces are in balance, so that there is no residual acceleration either along the isobars or perpendicular to them, the flow is called gradient wind flow and the equation which describes this kind of flow is called the gradient wind equation. Figures 8.2(a) and 8.2(b) illustrate the balance of forces in schematic form for the anticyclonic system with high pressure in the centre, and the cyclonic system with low pressure in the centre. In order to define the gradient wind equation we will assume that the isobars are circular around the centre and equally spaced. In real weather systems this condition may not be met exactly, but the approximation is sufficiently near for most practical calculations. Vigorous depressions are certainly observed to be almost circular.

Taking the anticyclonic case for the northern hemisphere first we have

$$\frac{V^2}{R} - \frac{1}{\rho}\frac{\partial p}{\partial R} = fV \tag{8.3}$$

where V is the wind speed around the isobars and R is a radial coordinate measured from the centre of the system. To be strict we must assign a sign to f which is positive in the northern hemisphere and negative in the southern hemisphere. We must also assign a sign to the horizontal curvature of the flow whose magnitude is $1/R$. In conformity with the conventions of vector calculus we take it to be positive to the left of the direction of motion and negative to the right in both hemispheres. Such a development will describe the sign of the motion around anticyclones and depressions in both hemispheres. We will return to this treatment later.

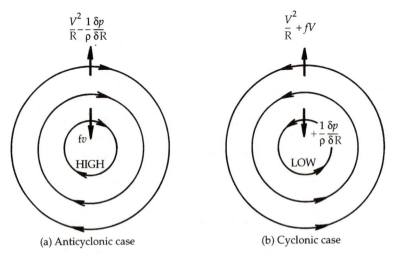

(a) Anticyclonic case   (b) Cyclonic case

FIGURE 8.2 Balance of forces for gradient wind flow in the northern hemisphere.

The cyclonic case for the northern hemisphere is

$$\frac{V^2}{R} + fV = -\frac{1}{\rho}\frac{\partial p}{\partial R} \tag{8.4}$$

Combining (8.3) and (8.4) we get

$$\frac{V^2}{R} \mp fV - \frac{1}{\rho}\frac{\partial p}{\partial R} = 0 \tag{8.5}$$

where the minus sign in the second term refers to the northern anticyclonic or southern cyclonic case and the plus sign refers to the northern cyclonic or southern anticyclonic case.

Equation (8.5) is a quadratic with solutions

$$V_{gr} = \frac{fR}{2} \mp \sqrt{\frac{f^2R^2}{4} - \frac{R}{\rho}\frac{\partial p}{\partial R}} \tag{8.6}$$

and

$$V_{gr} = -\frac{fR}{2} \pm \sqrt{\frac{f^2R^2}{4} + \frac{R}{\rho}\frac{\partial p}{\partial R}} \tag{8.7}$$

Equation (8.6) is valid for the northern anticyclonic or southern cyclonic case and (8.7) for the northern cyclonic or southern anticyclonic case. A word of explanation is needed here about signs. Since R is a radial coordinate, $(R/\rho) \cdot (\partial p/\partial R)$ is negative for anticyclones and positive for cyclones in both hemispheres.

We are interested in calculating the gradient wind which is caused by the pressure gradient. For such a relation the gradient wind will be zero if the pressure gradient is zero. We see at once that the latter condition is met in the northern hemisphere if we take the upper root in equations (8.6) and (8.7):

$$V_{gr} = \frac{fR}{2} - \sqrt{\frac{f^2R^2}{4} - \frac{R}{\rho}\frac{\partial p}{\partial R}} \tag{8.8}$$

and

$$V_{gr} = -\frac{fR}{2} + \sqrt{\frac{f^2R^2}{4} + \frac{R}{\rho}\frac{\partial p}{\partial R}} \tag{8.9}$$

We see now that if the pressure gradient term is zero the gradient wind is zero. In the southern hemisphere, $fR/2 < 0$, so that the lower roots must be chosen in (8.6) and (8.7) for the cyclonic and anticyclonic cases, respectively.

Equations (8.8) for the northern anticyclonic case and (8.9) for the northern cyclonic case and, of course, their southern hemisphere equivalent are most interesting and tell us a great deal about possible flows around circular isobaric systems.

8.3.1 Gradient wind solution for the anticyclonic case

We have already seen that equation (8.8) meets the condition that the gradient wind is zero when the pressure gradient is zero. The next thing that strikes us is the possibility of a negative sign under the square root radical in the anticyclonic case. This tells us that if the numerical value of the anticyclonic pressure gradient exceeds a certain limiting value we have an imaginary term. What does this mean? Mathematics is exciting in that it can on occasions tell us when physical laws are being contravened. The limiting value in equation (8.8) occurs when

$$\frac{f^2 R^2}{4} = \frac{R}{-\rho} \frac{\partial p}{\partial R}$$

that is, when the pressure gradient acceleration is greater than $f^2 R/4$. We then have for the northern anticyclonic case

$$V_{gr} = 0 \qquad \text{when} \quad \frac{\partial p}{\partial R} = 0$$

$$0 < V_{gr} < \frac{fR}{2} \qquad \text{when} \quad 0 < \frac{1}{-\rho} \frac{\partial p}{\partial R} < \frac{f^2 R}{4}$$

$$V_{gr} = \frac{fR}{2} \qquad \text{when} \quad \frac{1}{-\rho} \frac{\partial p}{\partial R} = \frac{f^2 R}{4}$$

$$V \text{ is imaginary} \quad \text{when} \quad \frac{1}{-\rho} \frac{\partial p}{\partial R} > \frac{f^2 R}{4}$$

and similarly for the southern hemisphere case.

The values of f and R are variable and thus determine whether the root of the quadratic is real or imaginary. The root is liable to become imaginary if

1. The pressure gradient becomes too large.
2. The radius of curvature of the high-pressure system becomes too small.
3. The latitude is near the equator, that is f becomes too small.

Thus if any or a combination of the three conditions above occur gradient wind flow cannot exist. What this means, physically, is that there is a component of the velocity which crosses the isobars. This has interesting connotations. It means that under such conditions an anticyclone must spread out, since part of the flow is blowing radially outwards from the centre across the isobars. It is not possible to have intense centres of high pressure with small radii, except perhaps for short periods of time. Nor is it possible to have anticyclonic systems at or near the equator. Any such systems would dissipate rapidly, on a time scale of hours. These restrictions on the horizontal structure of anticyclones are supported by observations which indicate that high-pressure systems are extensive and have light winds over their large central areas.

8.3.2 Gradient wind solution for the cyclonic case

We will now look at equation (8.9). We again see that this state meets the condition that the gradient wind is zero when the pressure gradient is zero. However, we see now that the expression under the radical must always be positive. There is therefore no limiting condition for which the expression becomes imaginary. Thus, there is theoretically no limit to the depth of a low-pressure centre. This feature is again borne out by observations. Depressions are generally smaller in size than anticyclones. The more violent they are, the smaller are the centres of low pressure. This is particularly observed in tropical cyclones, hurricanes and typhoons, all of which are different names for intense low-pressure systems which form in the tropics. The centres of such storms are called 'eyes'. They can be readily seen on satellite pictures, and are sometimes only a dozen or so kilometres in diameter. These properties are predicted by the equations we have derived.

8.4 THE CYCLOSTROPHIC WIND

We may set a balance between the centrifugal force and the pressure gradient force. This balance is only valid for the cyclonic case (*see* Fig. 8.2(b)), and only if we neglect the Coriolis force. We can do this if f is small, and/or if the time and space scales are small compared with the scale of normal synoptic systems seen on the weather map. We have

$$\frac{V^2}{R} = \frac{1}{\rho}\frac{\partial p}{\partial R}$$

$$V = \sqrt{\frac{R}{\rho}\frac{\partial p}{\partial R}}$$

(8.10)

Equation (8.10) describes the motion around small-scale systems such as tornadoes or waterspouts. It is theoretically possible for the flow to rotate in either an anticyclonic or a cyclonic sense, but is probably more frequent in the cyclonic sense.

8.5 THE INERTIAL WIND

We may also obtain a balance between the Coriolis force and the centrifugal force. This can only occur if there is no pressure gradient and the motion results from an initial impulse such as the firing of a projectile. (See problem 2 in Chapter 7 about the billiard ball.) It can only occur for clockwise flow in the northern hemisphere and anticlockwise flow in the southern hemisphere. Thus, $V^2/R = fV$ and $V = fR$.

8.6 THE 'STRANGE ROOTS' OF THE GRADIENT WIND EQUATION

Earlier in this chapter we defined the anticyclonic case as one in which we had a centre of high pressure, and a cyclonic case as one in which we had a low-pressure centre. In the northern hemisphere the wind normally blows clockwise around a high-pressure centre and anticlockwise around a low-pressure centre and vice versa in the southern hemisphere. Let us now consider the remaining roots of (8.6) and (8.7), namely

$$V_{gr} = +\frac{fR}{2} + \sqrt{\frac{f^2 R^2}{4} + \frac{\partial p}{d(\ln R)}} \qquad (8.11)$$

and

$$V_{gr} = -\frac{fR}{2} - \sqrt{\frac{f^2 R^2}{4} + \frac{\partial p}{d(\ln R)}} \qquad (8.12)$$

where (8.11) is the solution for the northern anticyclonic case and (8.12) is the solution for the northern cyclonic case for the condition that the flow is normal around the pressure centres. Note that the second term under the radical has been expressed in a different form since R is always positive.

Let us look at the 'strange' or anomalous roots which are those solutions given by the addition of the first term and the term under the radical. For the case (8.12), we have a non-zero value when the pressure gradient is zero. This is $V_{gr} = -fR$. We have already found that this is the inertial wind. It is negative and therefore anticyclonic and occurs when a body has been projected with some initial and constant speed V. The case of the anticyclonic 'strange' root has not been discussed in detail in most texts. It tends to be dismissed as an irrelevance. However, this root does have physical meaning. It is larger than the normal gradient wind and occurs as a result of the addition of an inertial component which an air parcel received at some earlier state and carries with it to a new location. Whereas geostrophic balance is a fairly stable state, gradient wind balance is very sensitive and only applies to some exact value of the radius of curvature. In practice it may be all right to neglect the 'strange' root, but we should know the physical meaning of the mathematical 'strangeness'.

There is another root for the northern hemisphere high-pressure case (8.11), but it does not meet the conditions we originally set up. This occurs when the flow around the anticyclone is anticlockwise in the northern hemisphere and clockwise in the southern hemisphere, that is 'the wrong way round'. This is an impossible condition for gradient wind balance as can be seen from Fig. 8.2.

The anomalous solution given for the low pressure case is obtained by adding the two negative terms in (8.12). The velocities are negative or anti-cyclonic. This situation is called antibaric because the flow is 'the wrong way round'. This type of flow cannot occur in normal-scale synoptic situations since it is the Coriolis force which determines the direction of rotation about the centre. It can only occur if the Coriolis force is small compared with the pressure gradient force. It is necessary for an inertial force to set up this kind of motion. This can happen in small-scale vortices such as tornadoes or waterspouts, or for a whirlpool down a bath plughole. The circulation is then cyclostrophic and defined by equation (8.10). One could imagine developing 'strange' cyclonic flow if a huge giant stirred up the atmosphere with an enormous teaspoon so that it generated the necessary speed for balance between the three forces to be established.

8.7 THE BALANCE EQUATION

There is a more complex equation which expresses a balance between forces for the general case of the gradient wind equation for isobars which change their curvature in space. This, of course, is what happens in the real world. Examination of any weather map will show isobars which form pressure patterns in space. Sometimes the patterns are in the shape of symmetrical waves, other times they are more complex. This equation is called the balance equation. The mathematical derivation of the equation will be postponed until a later chapter as it involves some new ideas we have not yet introduced.

8.8 PROBLEMS

1. A case of balanced motion might be that of a weather satellite which is positioned so that it remains permanently over the same spot on the equator. At what height must the satellite orbit? [Hint: Assume $g = 9.8/[1 + (z/a)^2]$ where z is the height above the surface of the earth of radius a.

2. What is the maximum possible gradient wind expressed as a multiple of the geostrophic wind for the same spacing of isobars for the regular anticyclonic case?

3. A circular shaped anticyclone has a pressure gradient of $1\,\text{hPa}\,\text{km}^{-1}$. What is the gradient wind at the following radii from the centre: (a) 10 km, (b) 100 km, (c) 500 km? Let $f = 10^{-4}\,\text{s}^{-1}$. Assume the density of the air is $1\,\text{kg}\,\text{m}^3$. What is the gradient wind for a cyclonic low-pressure system having the same pressure gradient at the same distances from the centre?

4. The funnel of a tornado which has a radius of 25 m rotates like a solid body at 1 revolution per second. What is the central pressure if the pressure at the funnel outer wall is 1000 hPa? Assume a temperature of 20°C. What is the velocity of the funnel wall?

5. In problem 4 assume that the velocity can be represented by the function $V = k/r^n$, where k is a constant. If the velocity decreases to $1\,\text{m}\,\text{s}^{-1}$ at a radius of 500 km, what is the value of the index n?

6. Consider the more rigorous form of the equations of motion in the tangent plane coordinate system. Neglecting terms involving vertical motion and residual accelerations show that there are two cases of balanced flow. If there is no zonal gradient but a meridional pressure gradient of $n\,\text{hPa}\,\text{km}^{-1}$ (average surface pressure and temperature) find the balanced zonal winds

at 45° latitude for the two cases. Why is one of them highly unlikely on planet earth?

7. Compute the space-averaged gradient wind between some inner radius r_0 and outer radius r for a circular anticyclone with a pressure gradient acceleration of $16.0 \times 10^{-5}\,\mathrm{m\,s^{-2}}$ and Coriolis parameter $f = 10^{-4}\,\mathrm{s^{-1}}$. Now assume the inner radius is the minimum for which gradient wind balance can occur and an outer radius of 500 km. (Note: This is a problem to challenge the mathematically minded.)

9

UNBALANCED FLOW

9.1 INTRODUCTION

We introduced the previous chapter by stating what we meant by balanced and unbalanced flow. We said that balanced flow covered motion which does not have any residual accelerations along the coordinate axes. Unbalanced flow is motion which does have residual accelerations along the coordinate axes. In this chapter we will study these accelerations.

9.2 THE AGEOSTROPHIC WIND

The difference between the actual wind and the geostrophic wind is called the ageostrophic wind. Another term used to denote this difference is the geostrophic departure.

Rewriting the equations of motion (7.4) and the geostrophic equations we have, as before,

$$\frac{du}{dt} - fv = -\frac{1}{\rho}\frac{\partial p}{\partial x} \qquad (9.1a)$$

$$\frac{dv}{dt} + fu = -\frac{1}{\rho}\frac{\partial p}{\partial y} \qquad (9.1b)$$

and

$$u_g = -\frac{1}{\rho f}\frac{\partial p}{\partial y} \qquad (9.2a)$$

$$v_g = \frac{1}{\rho f}\frac{\partial p}{\partial x} \qquad (9.2b)$$

From (9.1) and (9.2) we obtain

$$\frac{du}{dt} - fv = -fv_g$$

$$\frac{dv}{dt} + fu = fu_g$$

or

$$\frac{\mathrm{d}u}{\mathrm{d}t} = f(v - v_{\mathrm{g}}) \tag{9.3a}$$

$$\frac{\mathrm{d}v}{\mathrm{d}t} = -f(u - u_{\mathrm{g}}) \tag{9.3b}$$

There are many factors which cause the wind to depart from its geostrophic value. Strictly speaking gradient-balanced winds around anticyclones and cyclones are ageostrophic, although we normally mean that the flow is unbalanced in the sense that there are accelerations along and perpendicular to the isobars. Frictional drag, which we will discuss in a later chapter, is one factor which causes the wind to blow across the isobars. Imbalances can occur, even in straight, equally spaced isobars, if they are not orientated east–west, but have a meridional component. This is because the Coriolis force changes with latitude. But the most important geostrophic departures are normally associated with changes of the pressure gradient with time. To understand the physical meaning of this we will expand the so-called substantial derivative of pressure with time:

$$\frac{\mathrm{d}p}{\mathrm{d}t} = \frac{\partial p}{\partial t} + \frac{\partial p}{\partial x}\frac{\mathrm{d}x}{\mathrm{d}t} + \frac{\partial p}{\partial y}\frac{\mathrm{d}y}{\mathrm{d}t} \tag{9.4}$$

The term on the left hand side is the total derivative following the fluid. The first term on the right hand side is the local change and the other two terms on the right hand side are the advective changes. The local change means that the pressure is changing with time, as measured by a barometer which is located at a specific place. The changes given by the other terms are the changes which would be observed if we moved through a given pressure pattern on the weather map, carrying our barometer with us. If the local change is zero we say that the system is in a steady state. If the local change is not zero the system is in an unsteady state. Changes in $\partial p/\partial t$ are important because this is the term which can tell us whether a depression or cyclone is deepening, that is intensifying, or if an anticyclone is weakening or intensifying. Such changes in synoptic systems are known as development. In weather prediction it is most important to know if synoptic systems are developing, or if they are in a steady state. We will discuss an example of ageostrophic winds caused by development.

9.3 THE ISALLOBARIC WIND

We will assume a pattern of straight isobars. The geostrophic winds are given by equations (8.1). It is noted that in the geostrophic case the advective acceleration terms are zero since there is no change in the shape of the pattern in space:

$$u_{\mathrm{g}} = -\frac{1}{\rho f}\left(\frac{\partial p}{\partial y}\right)$$

$$v_{\mathrm{g}} = \frac{1}{\rho f}\left(\frac{\partial p}{\partial x}\right)$$

We will assume further that the pressure gradient is changing in time at a constant rate, but that its shape remains the same, that is, the isobars remain straight. Then

$$\frac{\partial u_g}{\partial t} = -\frac{1}{\rho f}\frac{\partial \dot{p}}{\partial y} \tag{9.5a}$$

$$\frac{\partial v_g}{\partial t} = \frac{1}{\rho f}\frac{\partial \dot{p}}{\partial x} \tag{9.5b}$$

From equations (9.3) and (9.5) we may write

$$u' = -\frac{1}{\rho f^2}\frac{\partial \dot{p}}{\partial x} \tag{9.6a}$$

$$v' = -\frac{1}{\rho f^2}\frac{\partial \dot{p}}{\partial y} \tag{9.6b}$$

where u' and v' are the geostrophic departures and $\dot{p} = \partial p/\partial t$. We also assume that f is constant. In equations (9.6) the geostrophic departures are also called isallobaric winds. They are produced by the condition that the pressure gradient is changing with time. In other words, development of the pressure pattern is occurring. Isopleths or lines of equal rate of change of pressure at a given location are called isallobars and the magnitude of the isallobaric wind can be calculated from the spacing of the isallobars in a way similar to that used to find the geostrophic wind from the spacing of the isobars. Figure 9.1 shows an example of the isallobaric wind for a case where an isallobaric high is super-imposed upon a system of straight isobars. Equations (9.6) are derived for a

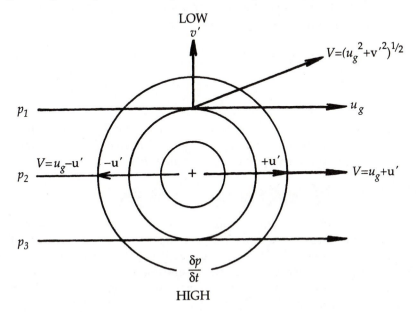

FIGURE 9.1 The isallobaric wind with westerly geostrophic flow, u_g. The actual wind is denoted by V_g and the isallobaric by u', v' (northern hemisphere).

given point where the rate of change of pressure gradient with time is constant. The isallobaric winds in Fig. 9.1 would vary according to the value of the isallobars at different points on the diagram. We might also consider cases where isallobaric centres were superimposed over circular anticyclonic or cyclonic systems. Thus an isallobaric high superimposed over an anticyclone would cause an isallobaric wind to flow outwards all round the anticyclone. The opposite effect would occur if an isallobaric low were superimposed over a cyclone or centre of low-pressure. Similarly, an isallobaric low over an anticyclone would cause an isallobaric wind to flow inwards towards the centre.

In the more general case we would expand the acceleration terms in the the equations of motion

$$\frac{du}{dt} = \frac{\partial u}{\partial t} + u\frac{\partial u}{\partial x} + v\frac{\partial u}{\partial y} \tag{9.7a}$$

$$\frac{dv}{dt} = \frac{\partial v}{\partial t} + u\frac{\partial v}{\partial x} + v\frac{\partial v}{\partial y} \tag{9.7b}$$

In the case of the isallobaric wind discussed above the advective terms are zero since there is no change of the pressure field in space. If the advective terms are not zero they contribute to the ageostrophic wind. In the simpler case of circular isobars which are not changing with time the ageostrophic wind is $\pm V^2/r$. Note that it has come to be accepted that the isallobaric wind is caused by the change of the pressure gradient with time, whereas all winds which are not geostrophic are strictly ageostrophic. It should be emphasized that the flow can be balanced, such as in the case of the gradient wind, but still be ageostrophic. If we considered the case where the intensity of a high- or a low-pressure system with circular isobars was changing with time we would have

$$\frac{2V}{R}\frac{\partial V}{\partial t} \mp f\frac{\partial V}{\partial t} = +\frac{1}{\rho}\frac{\partial \dot{p}}{\partial R} \tag{9.8a}$$

$$V' = +\frac{1}{\rho f}\frac{\partial \dot{p}}{\partial R}\bigg/\left(\frac{2V}{R} \mp f\right) \tag{9.8b}$$

In equations (9.8) V' is the isallobaric wind resulting from the change with time of the gradient wind, which is itself ageostrophic. Note that if the curvature term in the denominator is zero equations (9.8) reduce to (9.6).

9.4 PRESSURE CHANGES

In weather prediction it might be considered that the primary task is to predict changes in the pressure pattern. The pressure pattern is closely related to weather. Anticyclones give rise to fine weather, particularly in summer in middle latitudes, and most of the time in subtropical latitudes. This is due to subsidence or slow sinking of the air mass which warms adiabatically in accordance with equation (2.43). Similarly, low-pressure areas give rise to cloudy and rainy weather because the air rises and cools adiabatically. It is most important therefore to be able to know the distribution of highs and lows on a weather map. To do this we need to look more closely at equation (9.4). There are two problems to consider. The first is the development term expressed by $\partial p/\partial t$, and the second is the advective process given by the remaining terms. We will deal with the advective term first as this is the simpler process. We shall see in a later chapter that highs and lows are effectively advected, literally blown along by the upper wind flow. We can therefore estimate the upper air velocity by means of the geostrophic assumption and calculate the speed and direction of the movements of the highs and lows across the weather map. We can also examine the isallobaric pattern. Anticyclones will tend to move parallel to a line connecting an isallobaric low to an isallobaric high, and vice versa for a depression. The latter method was first used by Sverre Petterssen, a famous Norwegian meteorologist who helped to prepare the D-Day landing forecast for the Allies in June 1944. A less exact method is to extrapolate the tracks taken by the highs and lows during the past 24 hours. These methods are empirical, that is they rely on the movements remaining constant up until the time for which the forecast is required. Such methods were entirely relied upon in the past but in recent years numerical prediction models based on the full equations of motion, together with the thermodynamic equations predict the movements much more accurately, using small time steps of a minute or so. After every time step the forecast values for that time are used as initial conditions for the next time step, and so on. Although weather forecasters may use the old methods for local forecasts, they have the numerical forecasts besides as guides to help them.

We may express the advective terms of (9.4) more briefly in their vector form $\mathbf{V}\cdot\nabla p$. The term measures two kinds of changes, depending on whether the observer is stationary and the weather patterns are moving, or whether the observer is moving and the weather patterns are stationary. Thus, a barometer or barograph installed at a fixed location will show a trace which moves up and down the graph as the highs and lows pass over it. Alternatively, a barometer or barograph on a ship will fall if the ship is moving towards a cyclone, and rise if it is sailing towards an anticyclone. Motion is relative to the observer.

In this chapter we will be mainly concerned with the first term on the right-hand side of (9.4), which is known as the local rate of change of pressure. In general the latter is not exactly a measure of the intensification or weakening of a pressure system, because this is measured by the time rate of change of pressure at a point moving with the system. However, when a system is stationary, at the centre of a system or where a system is moving parallel to the isobars, the local rate of change of pressure does measure intensification, and it is therefore very useful. Often, we are specially concerned with pressure changes at the centre of a depression, particularly when the pressure there is falling (a synoptic term to mean decreasing). Synoptic meteorologists say the depression is deepening or filling according to whether the barometer is falling or rising at its centre. But what causes the local rises and falls of pressure which are observed continually on a weather chart? It is these continual changes which are responsible for the changing synoptic pressure patterns, and so for the weather which is associated with those patterns.

We know from the hydrostatic equation (5.1) that

$$p = \int_0^\infty \rho g \, dz \qquad (9.9)$$

Thus, the pressure shown by a barometer is just the weight of the entire air column above it. Changes in that pressure from one hour to the next result from changes which occur in that whole column. Where do these changes occur? At what levels in the atmosphere? We cannot answer these questions from the change in surface pressure, known as the surface pressure tendency, itself. There are, no doubt, various accumulations and depletions of the total mass of air occupying the different elements of volume of a column of air, but these usually cancel out to a large extent, leaving a small, residual net mass change which appears as a change in surface pressure. These accumulations of air within various elements of volume of the air column occur where there is convergence of the flow, and depletions of air occur within elements of volume where there is divergence of the flow. We will define convergence and divergence in mathematical terms in the next section. The observed changes of surface pressure do not result from a small depletion or accumulation of air which is occurring uniformly throughout the whole vertical column, but from the net residual of different magnitudes of convergence and divergence occurring at different levels throughout the vertical cross-section. We will see shortly that the accurate measurement and calculation of convergence and divergence is extremely difficult, if not impossible, by conventional means. In consequence it is very difficult to predict changes of surface pressure resulting from the development term, unless the forecaster has access to the output of a complex 'state-of-the-art' numerical model output. Even then, the numerical predictions do not always get it right.

To summarize, the observed pressure tendency is dependent on the integrated motion of the atmosphere from the surface upward to a level where pressure becomes inappreciable. Motion at some levels may be more important than others but it is the net result which matters.

9.5 DIVERGENCE AND CONVERGENCE

It has already been stated that convergence and its converse, divergence, represent an increase or decrease of mass within a specified volume. Convergence within a cross-section of a unit atmospheric column will cause a rise in pressure at the base of the cross-section considered. Likewise divergence within a similar cross-section will cause a fall of pressure at the base of the cross-section. These concepts are fundamental to dynamic and synoptic meteorology.

Let us suppose a rectangular box (Fig. 9.2) with faces ABCD and EFGH normal to the x axis, faces AEHD and BFGC normal to the y axis and faces AEFB and DHGC normal to the z axis.

Consider first the flow along the x axis. The mass of air entering face ABCD is $\rho u \, dy \, dz \, dt$. The mass of air leaving the face EFGH is

$$\left(\rho u + \frac{\partial(\rho u)}{\partial x} dx \right) dy \, dz \, dt$$

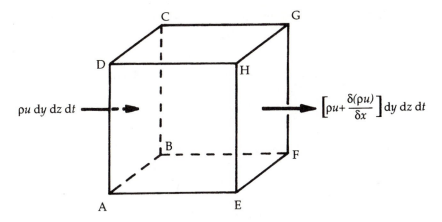

FIGURE 9.2 Derivation of the continuity equation.

The difference between the mass of air leaving this box and that entering it represents the increase or decrease of mass within the box.

Thus the difference may be written (mass going in less mass going out)

$$\Delta M = \rho u \, dy \, dz \, dt - \rho u \, dy \, dz \, dt - \frac{\partial(\rho u)}{\partial x} \, dx \, dy \, dz \, dt$$

$$\Delta M = -\frac{\partial(\rho u)}{\partial x} \, dV \, dt$$

where dV is an element of volume.

Similarly the differences resulting from the flow along the y and z axes are

$$-\frac{\partial(\rho v)}{\partial y} \, dV \, dt \quad \text{and} \quad -\frac{\partial(\rho w)}{\partial z} \, dV \, dt$$

respectively. The total difference is therefore

$$\left[-\frac{\partial(\rho u)}{\partial x} - \frac{\partial(\rho v)}{\partial y} - \frac{\partial(\rho w)}{\partial z} \right] dV \, dt \tag{9.10}$$

If (9.10) is divided through by $dV \, dt$ the resulting expression represents the change in mass from unit volume in unit time.

Now any change in the mass occupying the box considered must result in a change in the density of the mass contained therein. This density will have changed from ρ to $\rho + \partial \rho / \partial t$ in time dt. The change in density will accordingly be $(\partial \rho / \partial t) \, dt$ or $\partial \rho / \partial t$ in unit time. We may obviously equate the change in density with the advective change in mass:

$$\frac{\partial \rho}{\partial t} = -\frac{\partial(\rho u)}{\partial x} - \frac{\partial(\rho v)}{\partial y} - \frac{\partial(\rho w)}{\partial z} \tag{9.11}$$

or

$$\frac{\partial \rho}{\partial t} + \frac{\partial(\rho u)}{\partial x} + \frac{\partial(\rho v)}{\partial y} + \frac{\partial(\rho w)}{\partial z} = 0 \tag{9.12}$$

The equation (9.12) is called the equation of continuity. The expression

$$\frac{\partial(\rho u)}{\partial x} + \frac{\partial(\rho v)}{\partial y} + \frac{\partial(\rho w)}{\partial z}$$

is known as the mass divergence. It represents the loss of mass in the rectangular box in Fig. 9.2. We may therefore write

$$\text{div}(\rho \mathbf{V}) = \frac{\partial(\rho u)}{\partial x} + \frac{\partial(\rho v)}{\partial y} + \frac{\partial(\rho w)}{\partial z} \tag{9.13}$$

where $\mathbf{V}$ is the vector velocity. We note that whenever we write the quantity divergence in the form div $\mathbf{V}$ we must express the velocity as a vector. This is because divergence is a vector operator $\mathbf{i}(\partial/\partial x) + \mathbf{j}(\partial/\partial y)$ operating on the vector wind. The product of the operation is a scalar.

If purely horizontal flow is considered (9.13) takes the form

$$\text{div}_H(\rho \mathbf{V}) = \frac{\partial(\rho u)}{\partial x} + \frac{\partial(\rho v)}{\partial y} \tag{9.14}$$

It may be assumed normally that changes in density are small compared with changes in velocity, for horizontal flow. Then

$$\text{div}_H \mathbf{V} = \frac{\partial u}{\partial x} + \frac{\partial v}{\partial y} \tag{9.15}$$

Equation (9.15) represents the horizontal divergence of the velocity field.

Now let us consider a small slice cross-section of an air column in which any net increase or decrease of air caused by changes in the horizontal flow passes out through the vertical boundaries. The net divergence within the slice will in this case vanish, and

$$\frac{\partial(\rho u)}{\partial x} + \frac{\partial(\rho v)}{\partial y} + \frac{\partial(\rho w)}{\partial z} = 0$$

If the density does not change appreciably

$$\frac{\partial u}{\partial x} + \frac{\partial v}{\partial y} + \frac{\partial w}{\partial z} = 0$$

and

$$\frac{\partial u}{\partial x} + \frac{\partial v}{\partial y} = -\frac{\partial w}{\partial z}$$

or

$$\text{div}_H \mathbf{V} = -\frac{\partial w}{\partial z} \tag{9.16}$$

Equation (9.16) means that any horizontal divergence is compensated for by the removal or replacement of air by means of convergence of vertical motion and vice versa if there is to be no loss in mass. This is a fundamental mechanism of the working of the atmosphere and it helps us a great deal in formulating our ideas about weather forecasting. However, it is not exactly true or we would have no surface pressure changes. Clearly, vertical convergence does not always exactly balance horizontal divergence. Three-dimensional divergence in any layer may then be regarded as a small residual of the horizontal divergence which is not balanced by convergence in the vertical.

We may expand the total change in density $d\rho/dt$ and write

$$\frac{d\rho}{dt} = \frac{\partial\rho}{\partial t} + u\frac{\partial\rho}{\partial x} + v\frac{\partial\rho}{\partial y} + w\frac{\partial\rho}{\partial z}$$

$$\frac{\partial\rho}{\partial t} = \frac{d\rho}{dt} - u\frac{\partial\rho}{\partial x} - v\frac{\partial\rho}{\partial y} - w\frac{\partial\rho}{\partial z}$$

but from (9.12)

$$\frac{\partial\rho}{\partial t} = -u\frac{\partial\rho}{\partial x} - v\frac{\partial\rho}{\partial y} - w\frac{\partial\rho}{\partial z} - \rho\frac{\partial u}{\partial x} - \rho\frac{\partial v}{\partial y} - \rho\frac{\partial w}{\partial z}$$

Equating the two expressions above for $\partial\rho/\partial t$

$$\frac{\partial\rho}{\partial t} = -\rho\frac{\partial u}{\partial x} - \rho\frac{\partial v}{\partial y} - \rho\frac{\partial w}{\partial z} = -\rho\operatorname{div}\mathbf{V} \qquad (9.17a)$$

$$-\frac{1}{\rho}\frac{\partial\rho}{\partial t} = \operatorname{div}\mathbf{V} = \frac{1}{\alpha}\frac{\partial\alpha}{\partial t} \qquad (9.17b)$$

where α is the specific volume.

Figure 9.3 shows typical profiles of divergence and convergence within anticyclones and depressions. The profiles are schematic only and do not attempt to assess magnitudes. In an intensifying anticyclone upper level convergence tends to increase the central pressure. The isallobaric gradient at the surface causes an isallobaric wind which flows outwards (Fig. 9.3(a)). In response to the continuity equation there is downward motion, subsidence, adiabatic heating and fine weather, as described. In winter the subsiding motion may cause an inversion at the top of the boundary layer, sometimes called the mixing layer, at about 500–1000 metres. The sky may be totally covered by strato-cumulus cloud in such conditions, but the weather will be dry. This condition is sometimes called 'anticyclonic gloom'. Conversely, in a developing or deepening depression there is divergence in the upper levels. This results in an

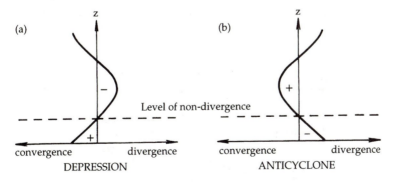

FIGURE 9.3 Typical convergence and divergence profiles over deepening depression and developing anticyclone.

isallobaric low at the surface (Fig. 9.3(b)). The isallobaric wind blows inwards and in response to the continuity equation there is an upward velocity of the air. The air cools in response to the adiabatic equation, reaches the condensation level, and rises further up the saturated adiabatic lapse rate. There is cloud and precipitation, sometimes gales, and generally unsettled and poor weather. Although for simplicity examples have been given for developing, that is intensifying, systems, the same profiles exist for steady state, and even decaying systems. An additional factor, that of surface friction, is always present which causes an acceleration across the isobars from high to low pressure, that is outwards around anticyclonic centres and inwards around cyclonic centres. The latter effect acts to re-enforce the effect of the isallobaric wind during the intensifying stages of the life of the system, but it may act in an opposite sense during the decaying stage. We will discuss the role of friction in a later chapter.

9.6 PRESSURE CHANGES IN GEOSTROPHIC FLOW

It has already been stated that a pressure change indicates a transfer of mass into or away from the atmospheric column at the base of which pressure is being measured. There can be no accumulation or depletion of mass if the flow is geostrophic. Geostrophic flow may vary in space and can be likened to a river which in places is wide and slow moving and in other places narrow and swift moving as through a gorge or canyon. The total amount of water passing a given cross-section across the river is everywhere the same. Imagine an isobaric pattern where the isobars are approaching one another (Fig. 9.4) and assume for the sake of argument that the flow is geostrophic. Consider that the flow is westerly so that the Coriolis parameter is constant, and that the density is constant. In the actual atmosphere the flow associated with a pattern such as that illustrated in Fig. 9.4 would not be exactly geostrophic. Flow which is not exactly geostrophic is called quasi-geostrophic. If, however, the flow in Fig. 9.4 is assumed exactly geostrophic the supposition is made that the velocity at every point between AB and DC adjusts itself instantaneously to the new value of the pressure gradient created by the approach of the isobars AD and BC towards one another.

Having made this assumption, consider the mass transfer within the area ABCD. From the geostrophic equation the velocity across the line AB is

$$V_1 = \frac{1}{\rho f} \frac{\delta p}{\delta n_1}$$

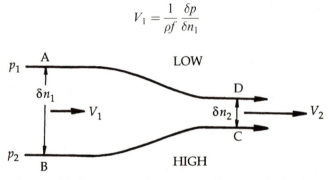

FIGURE 9.4 Geostrophic flow varying in space (northern hemisphere).

The velocity across the line DC is

$$V_2 = \frac{1}{\rho f} \frac{\delta p}{\delta n_2}$$

Then

$$\frac{\rho V_1}{\rho V_2} = \frac{\delta n_2}{\delta n_1}$$

and

$$\rho V_1 \delta n_1 = \rho V_2 \delta n_2 \tag{9.18}$$

Thus, if the flow is geostrophic the product of the velocity and the perpendicular distance between two isobars is proportional to the total mass transfer across that distance in unit time. Then equation (9.18) states that the mass transfer across AB is equal to the mass transfer across CD. If this is so there can be no accumulation or depletion of air between AB and CD, and consequently no change in pressure as measured between the base and top of the unit column of air considered. Obviously, if the flow is geostrophic and from the east or west, or nearly so, at all levels in the atmosphere div $\rho \mathbf{V}_g = 0$ and there is no pressure change at any level; therefore the pressure tendency measured at the surface of the earth must be zero. If the flow is from the north or south a correction must be made for the variation of the geostrophic speed due to the variation of f.

This truth that there can be no change of pressure if the flow is geostrophic can be derived more rigorously as follows.

Consider the hydrostatic equation (5.1) in the form $dp = -\rho g \, dz$ where p is the pressure at any level. We may integrate throughout a vertical column of the atmosphere with height of base z. Then $p = \int_z^\infty g\rho \, dz$; differentiating partially with respect to time

$$\frac{\partial p}{\partial t} = g \int_z^\infty \frac{\partial \rho}{\partial t} \, dz \tag{9.19}$$

Now, from the equation of continuity (9.12),

$$\frac{\partial \rho}{\partial t} + \frac{\partial(\rho u)}{\partial x} + \frac{\partial(\rho v)}{\partial y} + \frac{\partial(\rho w)}{\partial z} = 0$$

and substituting the expression for $\partial \rho / \partial t$ into (9.19) we have

$$\frac{\partial p}{\partial t} = -g \int_z^\infty \left(\frac{\partial(\rho u)}{\partial x} + \frac{\partial(\rho v)}{\partial y} \right) dz - g \int_z^\infty \frac{\partial(\rho w)}{\partial z} \, dz$$

$$\frac{\partial p}{\partial t} = -g \int_z^\infty \left(\frac{\partial(\rho u)}{\partial x} + \frac{\partial(\rho v)}{\partial y} \right) dz + g(\rho w)_z \tag{9.20}$$

Now, for geostrophic flow it follows from (8.1) that

$$\rho u_g = -\frac{1}{f} \frac{\partial p}{\partial y}$$

$$\rho v_g = \frac{1}{f} \frac{\partial p}{\partial x}$$

Differentiating the first of these equations with respect to x and the second with respect to y we obtain, if f is considered constant,

$$\frac{\partial(\rho u_g)}{\partial x} = -\frac{1}{f}\frac{\partial^2 p}{\partial x\,\partial y}$$

$$\frac{\partial(\rho v_g)}{\partial y} = \frac{1}{f}\frac{\partial^2 p}{\partial x\,\partial y}$$

Then, substituting the above expressions in (9.20)

$$\frac{\partial p}{\partial t} = -\frac{g}{f}\int_z^\infty \left(-\frac{\partial^2 p}{\partial x\,\partial y} + \frac{\partial^2 p}{\partial x\,\partial y}\right)dz + g(\rho w)_z$$

It is seen that

$$\frac{\partial p}{\partial t} = g(\rho w)_z \qquad (9.21)$$

If we integrate from a flat horizontal surface $z = 0$, w must be zero; thus

$$\frac{\partial p}{\partial t} = \frac{\partial p_0}{\partial t} = 0$$

and there can be no change of surface pressure with geostrophic flow. All pressure changes when the variation of f is negligible must therefore be due to ageostrophic motion.

If equation (9.20) is expanded,

$$\frac{\partial p}{\partial t} = -g\int_z^\infty \left[\left(u\frac{\partial \rho}{\partial x} + v\frac{\partial \rho}{\partial y}\right)dz + \left(\rho\frac{\partial u}{\partial x} + \rho\frac{\partial v}{\partial y}\right)dz\right] + g(\rho w)_z \qquad (9.22)$$

The first term on the right hand side of (9.22) represents the effect of the integrated horizontal advection of air of different density above height z on the pressure at height z. The second term is the effect on the pressure at z of the integrated horizontal divergence or convergence of velocity. The third term is the vertical motion term.

If the flow is strictly geostrophic and horizontal the vertical velocity term vanishes throughout and the advective and divergence terms balance one another. Thus from (9.21) and (9.22) $\partial p/\partial t = 0$ and

$$-g\int_z^\infty \left(u_g\frac{\partial \rho}{\partial x} + v_g\frac{\partial \rho}{\partial y}\right)dz = g\int_z^\infty \left(\rho\frac{\partial u_g}{\partial x} + \rho\frac{\partial v_g}{\partial y}\right)$$

$$\rho\left(\frac{\partial u_g}{\partial x} + \frac{\partial v_g}{\partial y}\right) = -u_g\frac{\partial \rho}{\partial x} - v_g\frac{\partial \rho}{\partial y}$$

and div $\rho\mathbf{V}_g = 0$ assuming constant density in the horizontal plane.

If the flow is predominantly northward or southward, the variation of f, the Coriolis parameter, will create divergence of the geostrophic wind.

9.7 MEASUREMENT OF DIVERGENCE

Divergence is clearly a quantity which it would be most useful to measure quantitatively. It is closely related to pressure change, a parameter which is linked to the problem of producing a forecast chart and so to the whole technique of weather forecasting. It has been shown in Section 9.6 that one cannot measure divergence if the flow is geostrophic. But since the geostrophic assumption is used fairly generally in the analysis of synoptic charts we find ourselves at an impasse. Divergence and convergence could perhaps be measured from actual wind observations. A method has, in fact, been devised to do this from a triangle of upper wind stations. The amount of air flowing into and out of such a triangle can be calculated. Any difference between the ingoing and outgoing values would represent horizontal mass divergence or convergence. These computations, however, are insufficiently accurate to be of real use because the natural variations of the wind over the area considered, together with the observational errors in the values of the wind direction and speed themselves, are both of a greater order of magnitude than the divergence values to be computed. In addition the network of upper wind stations is too sparsely spread to enable triangular areas of most useful size to be used, except in a few relatively small areas.

Such calculations in selected areas may, however, be of considerable interest for research purposes since they may give indications of levels where divergence and convergence are a maximum. If the results were integrated throughout an entire atmospheric column they would, in theory, give an approximation to the observed barometric tendency. At present such results would be of little use as an operational forecasting tool. In addition to the difficulty of making a reasonable assessment of the divergence itself at any level, the supreme difficulty of computing the surface pressure tendency lies in the fact that the latter is a small residual of divergence and convergence of a greater order of magnitude occurring at various levels throughout the atmospheric column (Section 9.4).

The magnitude of divergence on the synoptic scale is usually between about $10^{-5}\,\mathrm{s}^{-1}$ and $10^{-6}\,\mathrm{s}^{-1}$. One may realize the difficulty of measuring divergence if an idealized case is considered. Consider a narrow west-to-east strip. Let this strip be, say, 100 kilometres in length. Assume that at the western end of the strip the exact wind speed is $20\,\mathrm{m\,s}^{-1}$ while at the eastern end it is $21\,\mathrm{m\,s}^{-1}$ (Fig. 9.5).

Now

$$\mathrm{div}\,\mathbf{V} = \frac{\partial V}{\partial x}$$

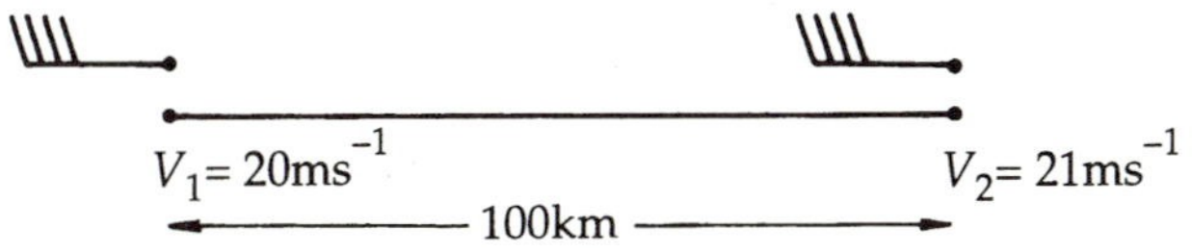

$$V_1 = 20\mathrm{ms}^{-1} \qquad V_2 = 21\mathrm{ms}^{-1}$$

$$\longleftarrow 100\mathrm{km} \longrightarrow$$

FIGURE 9.5 Calculation of divergence for one-dimensional flow.

If the derivative is replaced by a simple difference it follows that

$$\text{div } \mathbf{V} = \frac{V_2 - V_1}{L}$$

where L is the length of the strip.

If this condition extended throughout the whole depth of the atmosphere it can easily be shown that it would cause a fall of surface pressure of about 36 mb per hour throughout the area of the strip.

Now suppose we can only measure our wind speed at a given level to within 1 metre per second. The maximum possible error in the difference between the two wind speeds in the case quoted above would either increase the divergence to $3 \times 10^{-5}\,\text{s}^{-1}$ or else reverse the sign of divergence and give a value of $10^{-5}\,\text{s}^{-1}$ for convergence at that level. This would only be the value for one level. In effect there is direction to consider also, while at upper levels where the wind is stronger the accuracy of measurement is less. There we may only be able to measure the wind to within 5 metres or more per second. When such values are calculated for all levels and a net residual obtained it can be easily seen why it is so difficult to compute a realistic surface barometric tendency in this way.

We shall see in Chapter 11 that there is an alternative means of estimating the divergence, assuming that the flow is quasi-geostrophic, that is in more or less geostrophic balance. However, with the advent of high-speed computers the primitive equations, that is the full equations of motion, are used to compute pressure changes, as a residual of the divergence calculated for a number of upper levels. Certain tunings embodying the principles of the conservation of mass and energy are utilized to ensure that the predictions are realistic.

9.8 VERTICAL MOTION

In dynamic meteorology we are interested in large-scale vertical motion, that is on the scale of the synoptic chart. Vertical motion on this scale refers to the slow ascent or descent of air over comparatively extensive areas. The order of magnitude is centimetres per second; this is only about a hundredth part of the speed of horizontal motion. Vertical motion is responsible for most of our weather. Upward motion in the vicinity of depressions results in systems of cloud and precipitation while downward motion or subsidence in the vicinity of anticyclones causes clear skies. Large-scale vertical motion should not be confused with the scale of vertical motion associated with convective activity. The latter occurs over the much smaller areas covered by cumulus or cumulonimbus clouds and is not normally related to general rises or falls of pressure over a period of some hours. Convective vertical motion is, however, frequently of a magnitude which is comparable with or exceeds that of horizontal motion.

The equation of continuity expressed in (9.16)

$$\frac{\partial u}{\partial x} + \frac{\partial v}{\partial y} = -\frac{\partial w}{\partial z}$$

can be used to compute the vertical motion at different levels in the atmosphere, providing we have a vertical profile of the horizontal divergence throughout the atmospheric column. However, we must be wary of the accuracy of results achieved by this method over small areas, because of the errors which arise in the measurement of the horizontal divergence and the assumption of incompressibility inherent in the above form of the continuity equation. Large-scale numerical models, covering a large part of the earth's surface, are better equipped to deal with the problem, as they incorporate tuning and smoothing techniques as described at the end of the previous section.

9.9 PROBLEMS

1. Suppose an atmospheric column with a surface pressure of 1000 hPa was diverging horizontally throughout its entire length at $10^{-6} s^{-1}$. How long would it take for the surface pressure to reduce to its e-folding value? What is the e-folding value of the surface pressure?

2. Suppose an atmosphere is converging between the surface and 5 km at $10^{-5} s^{-1}$, and diverging between 5 km and 10 km at the same magnitude. Plot the vertical velocity profile. At what level does the vertical velocity reach a maximum? Discuss this result with reference to Fig. 9.4.

3. We have shown that the divergence of the geostrophic zonal wind is zero. What is the divergence of the meridional geostrophic wind?

4. In problem 9 at the end of Chapter 5 we found that if the earth's atmosphere was heated by 1 K the maximum pressure rise would occur at the height of the homogeneous atmosphere, assuming a constant lapse rate. How much horizontal convergence would be needed within a layer 100 hPa thick to produce the same result?

5. Compute the isallobaric wind for geostrophic flow at 45° latitude if the pressure gradient is 1 hPa per 100 km and the surface pressure is falling at 3 hPa per hour. Assume a density of $1.2 \, kg \, m^{-3}$. Compute the isallobaric wind for a circular-shaped depression for the same conditions at a radius of 100 km from the centre.

10

EULER AND LAGRANGE

10.1 INTRODUCTION

Euler stands on the station platform and measures the speed of the passing trains. Lagrange rides on an express train and measures its speed as it travels along the railway tracks, up and down inclines and through stations.

We may look at the motion of a fluid in two different ways. The first is the method attributed to Leonhard Euler who published his major work on material coordinates in 1755. The Eulerian technique involves observing the motion of parcels of matter from a fixed grid of points. Velocities are measured at intervals of time at each point on such a grid. The stationary observer measures changes in the properties of a fluid as the fluid streams by. In the Lagrangian technique the observer rides along with the parcels of matter and measures the motion and changes in properties of the parcels of fluid while in motion. There has been some discussion as to the relative roles played by these two great mathematicians (see *Weather Magazine* 1968, **23**, 2), but it is generally accepted that the two techniques were developed independently as described.

The Eulerian method is the one most commonly used in meteorology. Observations are made at specific locations. Fields of the variables observed may be smoothed and constructed in the forms of isopleths on a map. A grid array can then be placed over the fields and extrapolated values assigned to the array of points. Finite difference calculations can then be made over the grid space and substituted into the equations of motion in place of the partial derivatives. Thus, if we look again at equations (9.1), where the acceleration terms have been expanded in the form shown in (9.7), we would extract the values of u, v and f at the grid points, compute the spatial derivatives of u, v and p, and solve for $\partial u/\partial t$ and $\partial v/\partial t$. Having performed this calculation at every grid point on the chart for a given time interval, say a minute, we would repeat the operation for the next time interval, using the computed results as initial conditions for the ensuing computation. This operation could then be repeated for as long as desired. It is therefore clear that the Eulerian method is eminently suitable for numerical prediction models. However, the Lagrangian method does have some advantages over the Eulerian method for certain kinds of studies. The equations are often more amenable to integration and the scale of the motion is not fixed to a grid of fixed dimensions. The Lagrangian integrations will yield actual trajectories travelled by parcels of matter. The continuity equation can also be expressed in Lagrangian coordinates which follow the motion. This is very useful since divergence is not then fixed to the area or volume of a grid of constant size.

10.2 GEOSTROPHIC ADJUSTMENT: EXAMPLE OF THE LAGRANGIAN METHOD

Let us consider an illustrative example of the Lagrangian method and see what it tells us about the space and time scales which control the process of geostrophic adjustment, that is the process which forces the wind to blow along the isobars rather than across them. We will consider the equations of motion for frictionless flow where the isobars are oriented in an east–west direction, the pure geostrophic case:

$$\frac{du}{dt} - fv = -\frac{1}{\rho}\frac{\partial p}{\partial x} = P_x = 0$$

$$\frac{dv}{dt} + fu = -\frac{1}{\rho}\frac{\partial p}{\partial y} = P_y$$

(10.1)

where the pressure gradient term is assumed to remain constant. The first terms in the above equations now represent total derivatives following the motion, but we need not expand them as we did in (9.7). In this form (10.1) can be integrated. Solving the top equation for v, taking the derivative dv/dt and substituting in the second relation we have

$$\frac{d^2u}{dt^2} + f^2u = fP_y$$

(10.2)

We may simplify (10.2) by setting $m = uf - P_y$. Then we have $(d^2m/dt^2) + fm = 0$. The above is a second-order differential equation which has a type solution $m = A\sin ft + B\cos ft$. We may evaluate the constants at the initial starting point where $t = 0$, $u = u_0$, $v = v_0$. We have, since $u = (m + P_y)/f$, and $v = (1/f)(du/dt)$,

$$B = u_0 f - P_y, \quad A = fv_0$$

$$fu = v_0 f \sin ft + (fu_0 - P_y)\cos ft + P_y$$

Simplifying, we obtain

$$u = \left(\frac{u_0 - P_y}{f}\right)\cos ft + v_0 \sin ft + \frac{P_y}{f}$$

$$v = v_0 \cos ft - \left(\frac{u_0 - P_y}{f}\right)\sin ft$$

(10.3)

We shall consider the case where parcels start from rest to see how the geostrophic adjustment process works. Equations (10.3) then reduce to

$$u = \frac{P_y}{f}(1 - \cos ft)$$

$$v = \frac{P_y}{f}\sin ft$$

(10.4)

Equations (10.4) describe the velocity field as a function of time. Clearly the zonal velocity u reaches a maximum when $ft = \pi$, or $t = \pi/f$. At this time $u = 2P_y/f$ which will be recognized as just twice the geostrophic wind speed. At this time the wind is blowing along the isobars. It can also be seen from equation (10.4) that the zonal wind diminishes to zero at $t = 2\pi/f$. The latter time is called the inertial period. It is the order of magnitude of time needed for the geostrophic adjustment to take place. Table 10.1 shows the length of the inertial period for various latitudes. The inertial period is also called a half pendulum day, since it is 12 hours at the poles.

It is quite easy to integrate equations (10.4). The result will give us the trajectory of the parcel. Initiating the trajectory at the origin, we obtain

$$x = \frac{P_y}{f^2}(ft - \sin ft)$$

$$y = \frac{P_y}{f^2}(1 - \cos ft)$$

(10.5)

Equations (10.5) are equations of a cycloid, a well-known mathematical curve. Figure 10.1 shows a computer output of the track of a parcel following such a curve. It is assumed that the parcel starts from rest. A reasonable value of f is 10^{-4} which is valid for about $43°$ of latitude. A reasonable, although weak, pressure gradient is 1 mb (hPa) per $5°$ of latitude. Measuring pressure gradients by these criteria is sensible since maps are printed with latitude and longitude

TABLE 10.1 The inertial period at different latitudes

Latitude	Inertial period (h)
Equator	∞
5	$171.5 \approx 1$ week
10	69.1
15	46.4
20	35.1
25	28.4
30	24.0
45	17.0
60	13.8
90	12.0

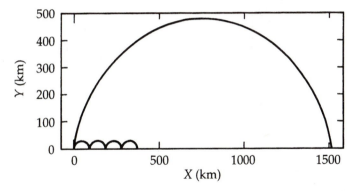

FIGURE 10.1 Numerical computation of parcel trajectories starting from rest for the geostrophic case. The pressure gradient force per unit mass is $15.0 \times 10^{-5} \, \mathrm{m \, s^{-2}}$. The upper curve corresponds to $10°$ North and the lower curve to $43°$ North.

lines for marking location. This magnitude of pressure gradient $P_y = 15.0 \times 10^{-5} \, \mathrm{m \, s^{-2}}$ is about right for normal surface pressure and temperature, say 1012 mb and 20°C. This value of the pressure gradient acceleration has been used to generate the curves in Fig. 10.1. Note that the amplitude and wavelength of the inertial wave are both 15 times larger at $10°$ than at $43°$ latitude, while the inertial period is four times longer.

In some non-mathematical descriptions of the geostrophic adjustment process, for the case described in this section, it has been said that the wind begins to flow across the isobars but eventually lines up along the isobars. At this stage geostrophic balance is achieved. This description only tells half the story. The simple mathematics tells us a great deal more of the detail of what must happen. Certainly, the geostrophic adjustment process depends on the length of the inertial period. In low latitudes where the inertial period is long, geostrophic adjustment is unlikely to occur because the pressure gradient will not stay constant for such long periods.

We can show that the geostrophic wind will result eventually, quite simply. The mean velocity of a parcel integrated over the inertial period is

$$\bar{u} = \frac{f}{2\pi} \left(\int u \, \mathrm{d}t \right) = \frac{f}{2\pi} \int \frac{\mathrm{d}x}{\mathrm{d}t} \, \mathrm{d}t = \frac{f}{2\pi} \int \mathrm{d}x \tag{10.6}$$

between limits of 0 and $(2\pi/f^2)P_y$, so that $\bar{u} = P_y/f$, which is the geostrophic wind speed. A similar operation shows that $\bar{v} = 0$. If we imagine now that we integrate over the paths of all parcels in a fluid we obtain the geostrophic velocity for the ensemble of parcels. In the first example shown in Fig. 10.2, the amplitude of the oscillation is only 30 km. The pressure gradient chosen is one that is small enough to occur in nature within the inertial period, so that it is reasonable to assume that geostrophic balance to a changing pressure gradient of this magnitude can and does occur for all practical purposes. In the second example shown in Fig. 10.2 geostrophic adjustment is made more difficult as the amplitude of the inertial oscillation is now 480 kms and the inertial period

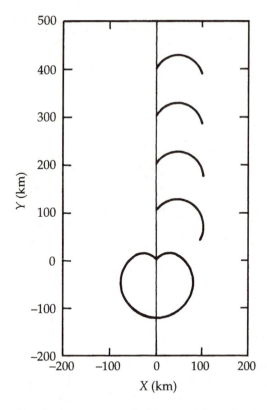

FIGURE 10.2 The anticyclonic case: numerical computation of trajectories of parcels starting from radii 0 to 400 km. Invariant pressure gradient of $15.0 \times 10^{-5}\,\mathrm{m\,s^{-2}}$, equivalent to about 1 mb (hPa) per degree of latitude, latitude of 10° North.

about 69 hours. However, in the tropics pressure gradient changes occur more slowly, so that there is more time for adjustment to occur. In some cases, if pressure gradient changes are not too rapid, quasi-geostrophic flow may occur at a latitude of 10° either side of the equator.

10.3 THE CASE OF THE ANTICYCLONE

We may use the method illustrated in the previous case to look at the case of an anticyclone with concentric circular isobars, once again holding the pressure gradient constant. This time we must use equations (10.5) in their polar coordinate form.

We then have

$$\ddot{r} - r\dot{\theta}^2 - fr\dot{\theta} = P_r$$

$$r\ddot{\theta} + 2\dot{r}\dot{\theta} + f\dot{r} = P_\theta = 0$$

(10.7)

where $P_\theta = -(1/\rho r)(\partial p/\partial \theta)$.

Integration of the second relation yields

$$r^2\dot{\theta} + \frac{fr^2}{2} = \text{constant} \tag{10.8}$$

If we assume that the air parcel starts from the centre of the anticyclone from a state of rest the constant of integration is zero and $\dot{\theta} = -f/2$. Substituting for $d\theta/dt$ in the first relation of (10.7) gives

$$\ddot{r} + \frac{f^2 r}{4} = P_r \tag{10.9}$$

We now follow the same procedure as for the pure geostrophic case and finally obtain

$$r = \left(\frac{4P_r}{f^2}\right)\left(1 - \cos\frac{ft}{2}\right) \tag{10.10}$$

Equation (10.10) will be recognized as that describing the curve of a cardioid. We note that the inertial period is double that for the geostrophic case while the amplitude is four times as large. Gradient wind adjustment is a more complicated process for circular isobaric systems than for the simple geostrophic case. Figure 10.2 shows the path of a parcel starting from radii of 0 to 400 kilometres from the centre. The tracks have been terminated at the time when they are tangential to the isobars.

In the anticyclonic case the families of trajectories which start from different radii, that is when r_0 is not zero, possess the interesting property that the minimum amplitude of the family occurs when $r = 3r_0$, measured from the origin of the coordinate system, that is from the centre of the anticyclone. For the case illustrated in Fig. 10.2 this occurs when $r = 85.3$ km. (See problem 3.)

10.4 THE CASE OF THE VARIABLE CORIOLIS PARAMETER

In the cases discussed in the previous section we have kept the Coriolis parameter constant during the calculations. This is a reasonable thing to do in temperate or polar latitudes, but not in tropical latitudes. As we have seen from Fig. 10.2 the trajectories have large amplitudes at 10° distant from the equator. The rate of change of the Coriolis parameter is a maximum at the equator and diminishes polewards since $\partial f/\partial y = (2\omega\cos\phi)/a$, where a is the radius of the earth. Thus, if we allow the Coriolis force to vary with latitude, using west–east-oriented isobars as for the geostrophic case, and let air parcels start from rest from different latitudes, say at 5° intervals, we find we have trajectories which possess some properties analogous to those illustrated in Fig. 10.3. Families of numerically computed trajectories of this kind are shown in Figs 10.4 and 10.5 for the northern hemisphere where the relation $f = \sin(5y - 40)$ was used, where $y = 8.0$ at the equator and increases/decreases by one unit for each 5° north or south of the equator. In Fig. 10.4 a pressure gradient of about 1 mb (hPa) per 5° of latitude is used. Note the convergence of the trajectories between those starting from the equator and those starting a few degrees away from the equator. The same relation applies for the variable Coriolis parameter for

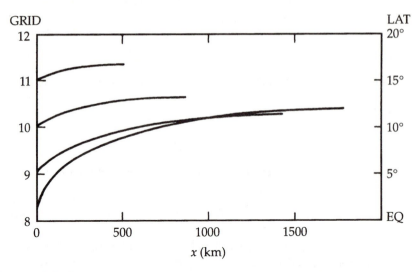

FIGURE 10.3 Trajectories of air parcels starting from rest at 5° intervals from the equator to 15° latitude in the northern hemisphere. The pressure gradient is assumed to be 1 mb (hPa) per 5° of latitude. This magnitude of pressure gradient might be found at upper levels of the winter troposphere in the tropics. Note the convergence of trajectories at about 11° latitude when they become parallel to the isobars. The grid on the left hand side is the function y where $y = \sin(5y - 40)$.

straight isobars as for the anticyclonic case with constant f, namely that for a given pressure gradient, the minimum latitude at which the wind becomes tangent to the isobars is just three times the starting latitude (problem 4).

Figures 10.4 and 10.5 give a simple, but interesting and useful, interpretation of the behaviour of the atmosphere in the tropics in the real world. In Fig 10.4 the gradient is small but of the order of magnitude which may exist in winter at upper levels polewards of the equator in one or the other hemisphere. Both the winter subtropical jet stream and the existence of the winter subtropical anticyclones can, at least partially, be explained by the pattern of trajectories appearing in this figure. Note the convergence of the trajectories where they become parallel to the isobars, particularly between 11° and 12° latitude. In Fig. 10.5 the pressure gradient is five times greater. This pattern is more applicable to the Indian summer monsoon when cross-equatorial pressure gradients occur from late June to September. Convergence of trajectories occurs now between about 19° and 21° latitude. In the last chapter we saw that upper level convergence tends to contribute to the building up of anticyclones. If Fig. 10.4 is a possible example of the behaviour of the troposphere in the tropics it suggests the presence of the subtropical anticyclones and subsidence in subtropical latitudes. On the other hand if Fig. 10.5 is used as an example to illustrate pressure gradients in the boundary layer across and polewards of the equator it suggests convergence and upward motion in subtropical latitudes. The main thrust of the Indian summer monsoon is observed in Bombay, which is about 19°N latitude. Observation of Bombay weather supports the theoretical conclusions derived from Fig. 10.5.

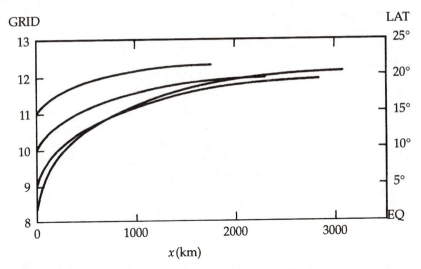

FIGURE 10.4 Same as Fig. 10.3, but with a pressure gradient of 1 mb (hPa) per degree of latitude, a value likely to be found in summer monsoon regions. Note the convergence of the trajectories at about 20° of latitude when they become parallel to the isobars.

The scientific method is to investigate whether observations in the real world confirm theoretical arguments. It may proceed in two ways. Perhaps the most exciting is to attempt to predict events suggested by the solutions of mathematical equations. The alternative is to collect series of observations and attempt to explain them in terms of solutions of equations. Either way there will probably

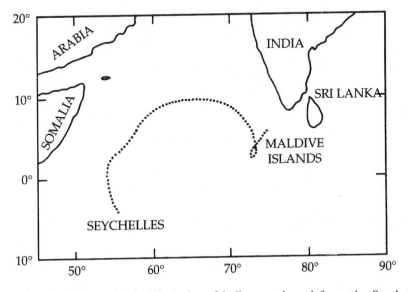

FIGURE 10.5 Constant-level trajectories of balloons released from the Seychelles (adapted from Cadet and Overlez, 1976).

be a need to refine and tune the theory, or model, so that it agrees more closely with real measurements. Einstein predicted from his theory of relativity that light rays would bend if they passed close to the sun. It needed a solar eclipse to prove his assertion.

Figure 10.5 shows the actual trajectories of balloons that were released from the Seychelles in the summer monsoon season of 1975 and tracked by satellite. All together 45 balloons were released so that they drifted at constant pressure within the tropical boundary layer. For practical purposes constant pressure is equivalent to constant height for the purpose of calculating the pressure gradient, as we shall see in a later chapter. It took many years before roughly computed tracks like those shown in Figs 10.3 and 10.4 could be confirmed by actual observations. The most interesting trajectory in Fig. 10.5 is the one marked with a dotted line, which shows a perfect inertial oscillation. It even shows a small loop, a characteristic predicted by the equations, depending on the initial conditions.

10.5 DIVERGENCE OF PARCELS IN A FLUID

If we consider a small moving slice of air of constant mass δM

$$\delta M = \rho A h$$

where A is the cross-section and h is the thickness of the slice considered. Then

$$\frac{d(\delta M)}{dt} = \frac{d(\rho A h)}{dt} = 0$$

$$\rho A \frac{dh}{dt} + \rho h \frac{dA}{dt} + A h \frac{d\rho}{dt} = 0$$

Dividing through by $\delta M = \rho A h$

$$\frac{1}{h}\frac{dh}{dt} + \frac{1}{A}\frac{dA}{dt} + \frac{1}{\rho}\frac{d\rho}{dt} = 0$$

$$\frac{-1}{\rho}\frac{d\rho}{dt} = \frac{1}{h}\frac{dh}{dt} + \frac{1}{A}\frac{dA}{dt} \tag{10.11}$$

We know from the continuity equation that

$$-\frac{1}{\rho}\frac{d\rho}{dt} = \text{div } \mathbf{V}$$

Hence

$$\frac{1}{h}\frac{dh}{dt} + \frac{1}{A}\frac{dA}{dt} = \text{div } \mathbf{V} \tag{10.12}$$

If the total divergence is zero (i.e., the flow is assumed to be incompressible), we obtain the relation

$$\frac{1}{A}\frac{dA}{dt} = -\frac{1}{h}\frac{dh}{dt} \tag{10.13}$$

In practice we interpret (10.13) to mean that if a slice of constant mass expands, its thickness contracts, and vice versa. We may put this concept into practice by following three balloons which are released from different locations and all ascend at the same rate. The rate of change of the area of the triangles formed by the three balloons is then a measure of the divergence (or convergence) of the slice of the atmosphere through which the balloons are ascending. It is sometimes useful to measure convergence in Lagrangian rather than Eulerian terms. For example, the Eulerian prediction of the convergence and therefore increased concentration of harmful pollutants within a square of constant area, the sides of which are 100 km, would be less useful than the Lagrangian prediction of the contraction of an area of pollutants to dimensions of, say, 1 km or less, when concentrations might be so high as to become lethal.

10.6 STREAMLINES

In this chapter we have been concerned with trajectories, the tracks followed by parcels of fluid (or individual bits of matter). A streamline is defined as a line joining points of tangency to the wind vector for every point of the flow. Thus $dy/dx = v(x, y, t)/u(x, y, t)$ states that the direction of the streamline at any point x, y coincides with the wind vector at a given time t_0. A streamline thus gives an instantaneous picture of the field of motion at a fixed time. Isobars represent streamlines if the flow is geostrophic or gradient and steady, that is if there is no change with time with reference to fixed points within the field of flow. Thus while the slope of a streamline, $dy/dx = v/u$, the slope of a trajectory is $dy/dx = v(x, y, t)/u(x, y, t)$, which is not restricted to a fixed time.

We may find a relation between trajectories and streamlines. If $d\beta$ expresses the change of direction of the wind, $r\, d\beta = ds$ where ds is an infinitesimal displacement along the horizontal trajectory and r is the radius of curvature. Then

$$r\frac{d\beta}{dt} = \frac{ds}{dt}$$

$$\frac{d\beta}{dt} = \frac{V}{r} = VK_t \tag{10.14}$$

where K_t is the curvature of the trajectory of the air. Now

$$\frac{d\beta}{dt} = \frac{\partial\beta}{\partial t} + u\frac{\partial\beta}{\partial x} + v\frac{\partial\beta}{\partial y} \tag{10.15}$$

where $\partial\beta/\partial t$ is the local turning of the wind.

However, $\partial\beta/\partial t = 0$ for a streamline by definition since the streamline refers to an instantaneous pattern of motion, so

$$u\frac{\partial\beta}{\partial x} + v\frac{\partial\beta}{\partial y} = V\frac{\partial\beta}{\partial s} = \frac{V}{r} = VK_s \tag{10.16}$$

where K_s is the curvature of the streamline. We may write from (10.14), (10.15) and (10.16)

$$\frac{\partial\beta}{\partial t} = V(K_t - K_s) \tag{10.17}$$

When the motion is steady

$$V(K_t - K_s) = 0$$

and

$$K_t = K_s$$

Trajectories and streamlines are then coincident.

10.7 THE STREAM FUNCTION

In later chapters we shall make use of a relationship called the stream function. We will introduce the definition here:

$$u = -\frac{\partial \psi}{\partial y}$$

$$v = \frac{\partial \psi}{\partial x}$$

$$(10.18)$$

We note that equations (10.18) are somewhat similar to the geostrophic equation. Like the geostrophic equation with a constant f, we easily find that

$$\frac{\partial u}{\partial x} + \frac{\partial v}{\partial y} = 0$$

Thus the divergence of any flow represented solely by a stream function is always zero. Non-divergent wind velocities can be represented by the spacing of isopleths of ψ on a map. The ensuing pattern gives a good visual interpretation of the non-divergent wind flow. A technique of streamline analysis is discussed in Chapter 16, but it should be noted that the streamlines described therein are not quite the isopleths mentioned above.

10.8 PROBLEMS

1. Derive equation (10.10) from equations (10.7).
2. At what radius does the tangential velocity of a parcel starting from rest from the centre of an anticyclone with concentric isobars attain maximum possible gradient wind speed for the conditions chosen? $f = 10^{-4}\,\mathrm{s}^{-1}$, pressure gradient acceleration $15 \times 10^{-5}\,\mathrm{m\,s^{-2}}$. At what angle does the wind intersect the isobars at this point?
3. In the example illustrated in Fig. 10.3 the pressure gradient force per unit mass is $15.0 \times 10^{-5}\,\mathrm{s}^{-2}$. What is the minimum distance from the origin for any trajectory to become parallel to the isobars? At what distance from the origin does this trajectory start? [Hint: Integrate equations (10.7) for the case where parcels start from $r = r_0$. Set the tangential velocity equal to zero and differentiate dr/dr_0.]
4. Show that for the case of a variable Coriolis parameter and isobars parallel to circles of latitude the minimum latitude at which the wind blows parallel to the isobars is three times the latitude from which the air parcel starts. What is this latitude for the pressure gradient value used in Fig. 10.3(a)?

[Hint: Substitute $f = 2\omega(y/a)$ into equations (7.4). This approximation is reasonable up to about $15°$ of latitude. Then integrate as in the previous question. At what latitude does a parcel become tangent to the isobars if it starts from the equator?]

5. Solve problem 3 using the equations of motion in spherical coordinates and hence show that the assumption $f = 2\omega(y/a)$ used in the previous problem is a fair approximation for tropical latitudes. [Hint: Answer is $\cos^2\phi = \cos^2\phi_0/[1 + P_y/(\sin 2\phi_0 a\omega^2)].$]

6. What is the slope of the cycloid derived in equations (10.5) at $t = \pi/2f$?

7. Obtain (10.11) from $\delta M = \rho A h$ by logarithmic differentiation.

VORTICITY

11.1 INTRODUCTION

We now come to an extremely important concept in dynamic meteorology. It is, in fact, a concept which is an integral part of fluid dynamics, the science concerned with the motion of fluids, gases and liquids. It is called vorticity and is a measure of rotation. Rotation is also sometimes called spin, a property that is a characteristic of quantum physics. Spin or rotation is therefore a fundamental property of our universe. We have already seen in earlier chapters that the wind blows around anticyclones and depressions. Such circulations have vorticity. The measure of the vorticity of the flow around these systems is a measure of the intensity of those systems. This chapter will discuss the derivation of vorticity as well as interpret its significance when applied to simple patterns on the weather map. It will be seen in this and later chapters that vorticity provides an extremely useful practical prediction tool.

11.2 CIRCULATION

The idea of circulation implies moving along a circular path, or more exactly, moving around a closed path. We talk about the circulation of money (although money spent may not always return to the spender). In the French language the word for traffic is *la circulation*. At parties the host or hostess will tell the guests to circulate. Although these descriptions may not be rigorous the idea is similar to the scientific one with which we shall here be concerned.

Mathematically,

$$C = \oint (u\,dx + v\,dy + w\,dz) \tag{11.1}$$

or, in a horizontal plane, $C = \oint (u\,dx + v\,dy)$.

We may also express the circulation as

$$C = \oint V \cos \alpha \, ds \tag{11.2}$$

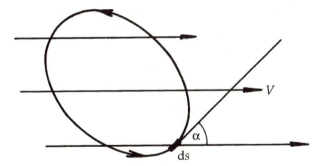

FIGURE 11.1 Circulation.

where V is the velocity of flow at any point in the field of motion and α is the angle between the direction of flow and the direction of the tangent to the element of distance along a closed path through the field of motion at that point, as shown schematically in Fig. 11.1.

In meteorology, where the geostrophic assumption is used in synoptic analysis, it is clear that the velocity field and the field of isobars are coincident and therefore that the circulation around a closed isobar may be expressed by $\oint V \, ds$. We must make the point here that when we refer to the geostrophic assumption as a description of the flow at any point in a field of motion, we really mean balanced flow. This could include gradient wind flow which, as we have seen, is more rigorous than geostrophic flow if the isobars are curved. But the pattern of isobars is far from being circular everywhere so that even the gradient wind equation is not exact. What we shall really mean by geostrophic flow is balanced flow in the sense that the wind blows parallel to the isobars at every point on the pressure pattern. There is an equation which expresses this condition which we shall come to later. For the time being, we may generalize and equate geostrophic flow with the quasi-geostrophic assumption that the flow is parallel to the isobars at every point and that its speed is approximately given by the geostrophic relation.

11.3 VORTICITY

Vorticity is a measure of rotation or spin. It may be defined as circulation per unit area, or as the quantity which, when integrated over the area, gives the circulation.

Thus

$$C = \int \zeta \, dA \tag{11.3}$$

or

$$dC = \zeta \, dA \tag{11.4}$$

where ζ is the vorticity about the axis normal to dA.

Consider a rotating disc of radius r. From (11.2)

$$C = \oint V \, ds \cos \alpha = \oint V \, ds = 2\pi r V$$

Here again $\alpha = 0$ since the motion is always around the closed path.

If we consider a small but finite rotating disc $\zeta = dC/dA$,

$$\zeta = \frac{2\pi r V}{\pi r^2} = \frac{2V}{r} = 2\omega$$

where ω is the angular velocity of the disc.

Now consider a larger rotating disc which is divided up into infinitesimal squares (Fig. 11.2). If the circulation around these squares is added up it is seen that adjoining sides supply equal and opposite contributions and cancel out, leaving the circulation around the perimeter.

Thus the vorticity of the rotating disc is 2ω where C is now the circulation around the perimeter of the disc and A is the whole area enclosed by the perimeter:

$$\zeta = 2\omega \tag{11.5}$$

The above relation then states that the vorticity of a rotating disc equals twice its angular velocity.

Since vorticity can be measured in terms of rotation or angular velocity it is clearly a function of the curvature of the path around which the parcel or element travels. This is particularly significant in meteorology where, as a result of the geostrophic assumption, the wind is considered to blow along the isobars. The curvature of the isobars is therefore of prime importance in estimating the horizontal vorticity of the wind field in the atmosphere. Vorticity is also a function of another parameter, namely shear. Shear is defined as the rate of change of velocity in a direction normal to the direction of motion. Thus, in

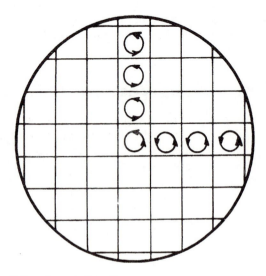

FIGURE 11.2 Vorticity of a rotating disc.

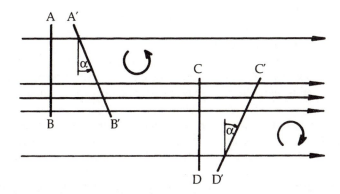

FIGURE 11.3 Cyclonic and anticyclonic shear in a westerly current (northern hemisphere).

meteorology, horizontal shear exists if, for example, the pressure gradient changes in a direction along the gradient, that is across the isobars.

Figures 11.3 and 11.4 show the generation of shear vorticity in westerly and easterly air currents. We might imagine momentarily that Figs 11.3 and 11.4 represent a flowing stream. The velocity of the current is inversely proportional to the distance between the streamlines, which we can also imagine to be isobars. If we place a stick across the stream we will observe that the stick rotates in the direction indicated. The portion of the stick lying in the stronger part of the current will move faster than the portion lying in the weaker part of the current. The stick will be seen to have rotated an angle α which the new position A'B' makes with the initial position AB and similarly with stick CD. In the northern hemisphere, cyclonic shear occurs where the rotation is anticlockwise, and anticyclonic shear occurs where the rotation is clockwise. We see that rotation has been generated through shear, even though the current itself does not possess any curvature in its path. In other words, we see that vorticity has been generated by the shear. It is shear vorticity as distinguished from curvature vorticity.

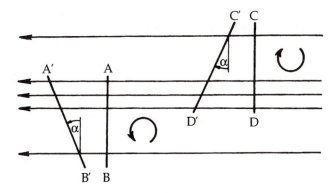

FIGURE 11.4 Cyclonic and anticyclonic shear in an easterly current (northern hemisphere).

Mathematically, shear is denoted by $\partial V / \partial r$ where the numerator is a small change in velocity while the denominator is a small leftward distance perpendicular to the direction of motion.

Figures 11.3 and 11.4 may refer to either hemisphere. In the northern hemisphere anticlockwise rotation is cyclonic and clockwise rotation anticyclonic. In the southern hemisphere the reverse is true. Remember that f changes sign and becomes negative in the southern hemisphere. However, the conventions of vector calculus require the normal coordinate to increase towards the left of the flow in both hemispheres, so cyclonic vorticity is positive (negative) and anticyclonic vorticity is negative (positive) in the northern (southern) hemisphere.

11.4 DERIVATION OF EXPRESSIONS FOR VORTICITY

We have seen in the previous section that vorticity is a function of the two parameters curvature and shear. We will now derive an expression for vorticity in polar coordinates which brings out the dual character of vorticity very nicely.

In Fig. 11.5 BA and CD are streamlines having a common centre of curvature O. Consider the circulation around an element of area ABCD where the velocity V is everywhere normal to the radius vector.

Starting at A we proceed cyclonically around the circuit ABCD. From (11.2)

$$dC = -Vr\,d\theta + 0 + \left(V + \frac{\partial V}{\partial r}dr\right)(r + dr)\,d\theta + 0$$

$$= -Vr\,d\theta + Vr\,d\theta + V\,dr\,d\theta + r\frac{\partial V}{\partial r}dr\,d\theta + \frac{\partial V}{\partial r}dr^2\,d\theta$$

The last term may be neglected since it is a differential of third order.

The area of the element is

$$dA = r\,dr\,d\theta$$

$$\zeta = \frac{dC}{dA} = \frac{V\,dr\,d\theta + r(\partial V / \partial r)\,dr\,d\theta}{r\,dr\,d\theta} = \frac{V}{r} + \frac{\partial V}{\partial r} \tag{11.6}$$

The first term represents curvature. The second term represents the change of velocity along r and is therefore a shear term. As previously stated we see that vorticity is composed of two components, curvature vorticity and shear

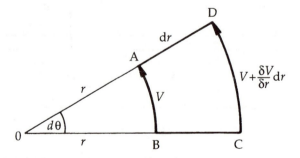

FIGURE 11.5. Derivation of vorticity in polar coordinates.

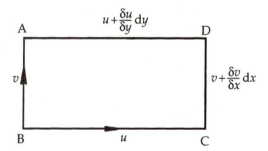

FIGURE 11.6 Derivation of vorticity in Cartesian coordinates.

vorticity. Note that though the positive (leftward) curvature of the flow is $1/r$, the normal coordinate itself is $-r$ in this case.

It is useful to derive an expression for vorticity in Cartesian coordinates also. To do this we consider the circulation around a small rectangular element in the same way as for the small element of area in deriving the expression (11.6) in polar coordinates. In this case we also proceed around the sides ABCD (Fig. 11.6).

Since in a horizontal plane

$$C = \oint (u\,dx + v\,dy)$$

we therefore have

$$dC = -v\,dy + u\,dx + \left(v + \frac{\partial v}{\partial x}dx\right)dy - \left(u + \frac{\partial u}{\partial y}dy\right)dx$$

$$= -v\,dy + u\,dx + v\,dy + \frac{\partial v}{\partial x}dx\,dy - u\,dx - \frac{\partial u}{\partial y}dx\,dy$$

$$= \left(\frac{\partial v}{\partial x} - \frac{\partial u}{\partial y}\right)dx\,dy = \left(\frac{\partial v}{\partial x} - \frac{\partial u}{\partial y}\right)dA$$

where dA is the area of the rectangular element, and

$$\zeta = \frac{dC}{dA} = \frac{\partial v}{\partial x} - \frac{\partial u}{\partial y} \qquad (11.7)$$

This is the vorticity perpendicular to the x, y plane. Vorticity perpendicular to the x, z plane is derived similarly and denoted by

$$\eta = \frac{\partial u}{\partial z} - \frac{\partial w}{\partial x}$$

Vorticity perpendicular to the y, z plane is denoted by

$$\xi = \frac{\partial w}{\partial y} - \frac{\partial v}{\partial z}.$$

11.5 RELATIVE AND ABSOLUTE VORTICITY

The three components of vorticity derived in the preceding section are created by the combined curvature and shear vorticity of the wind field. In this work we shall only be concerned with the first component, ζ, which is perpendicular to the x, y plane. This vorticity is called relative vorticity to indicate that it is relative to the surface of the earth. However, since the earth rotates it possesses its own vorticity, denoted by f, the Coriolis parameter. We proved this earlier when we showed that the vorticity of a solid rotating disc was twice its angular velocity. Thus every parcel of air possesses its own vorticity plus earth vorticity f, corresponding to the latitude about a z axis normal to the surface of the earth. The sum of the relative and earth vorticity is

$$\zeta_A = \zeta + f \qquad (11.8)$$

The sum is called the vertical component of absolute vorticity.

In the next section we shall be particularly concerned with the physical significance of rates of change of the absolute vorticity.

11.6 THE DIVERGENCE–VORTICITY RELATION

We now come to a remarkable and fundamental relation. We have seen that convergence and divergence of mass result in pressure changes, and that if we conserve mass in a vertical column the continuity equation gives a vertical velocity profile. The calculation of pressure changes gives us a means of predicting pressure patterns on the weather map. Vertical velocity profiles may be used to predict fine or rainy weather from the laws of thermodynamics discussed in Chapters 1 to 5. We have also come to the conclusion that convergence and divergence are quantities which are difficult to measure accurately by observational means. We have seen that we cannot calculate these elusive parameters from the weather map itself, using the geostrophic assumption, since the divergence of the geostrophic wind is zero for scales less than planetary. Although nowadays fast electronic computers use the full primitive equations of motion, this was not the case in the near past. Meteorology, both before and after World War II, was obsessed with the field of barometric pressure. This is not to say that pressure is not an important observation. It always has been and still is the main consideration in forecasting each day's weather. However, theoretically, there are other means of looking at the motion of the atmosphere, which can be applied in a very useful manner, particularly to the free atmosphere above the boundary layer. The breakthrough in this regard was the application of the vorticity theorem, which we will now derive.

In the horizontal x, y plane

$$\mathrm{div}_H \mathbf{V} = \frac{\partial u}{\partial x} + \frac{\partial v}{\partial y}$$

and perpendicular to it

$$\zeta = \frac{\partial v}{\partial x} - \frac{\partial u}{\partial y}$$

We may consider once more the equations of horizontal motion

$$\frac{du}{dt} - fv = -\frac{1}{\rho}\frac{\partial p}{\partial x}$$

$$\frac{dv}{dt} + fu = -\frac{1}{\rho}\frac{\partial p}{\partial y}$$

(11.9)

and differentiate the first of these above equations partially with respect to y and the second with respect to x.

Then,

$$\frac{\partial(du/dt)}{\partial y} - \frac{\partial(fv)}{\partial y} = -\frac{\partial((1/\rho)(\partial p/\partial x))}{\partial y}$$

$$\frac{\partial(dv/dt)}{\partial x} + \frac{\partial(fu)}{\partial x} = -\frac{\partial((1/\rho)(\partial p/\partial y))}{\partial x}$$

The Boussinesq approximation will be made that ρ remains constant in the horizontal. Then subtracting the second equation from the first

$$\frac{\partial(du/dt)}{\partial y} - \frac{\partial(dv/dt)}{\partial x} = \frac{\partial(fv)}{\partial y} + \frac{\partial(fu)}{\partial x} - \frac{1}{\rho}\frac{\partial^2 p}{\partial x \partial y} + \frac{1}{\rho}\frac{\partial^2 p}{\partial x \partial y}$$

Expanding the above and eliminating the pressure term

$$\frac{\partial}{\partial y}\left(\frac{\partial u}{\partial t} + u\frac{\partial u}{\partial x} + v\frac{\partial u}{\partial y}\right) - \frac{\partial}{\partial x}\left(\frac{\partial v}{\partial t} + u\frac{\partial v}{\partial x} + v\frac{\partial u}{\partial y}\right) = f\frac{\partial v}{\partial y} + v\frac{\partial f}{\partial y} + f\frac{\partial u}{\partial x}$$

It is noted that the Coriolis parameter is a function of latitude only so that

$$u\frac{\partial f}{\partial x} = 0$$

Then

$$u\frac{\partial^2 u}{\partial x \partial y} + \frac{\partial u}{\partial x}\frac{\partial u}{\partial y} + v\frac{\partial^2 u}{\partial y^2} + \frac{\partial u}{\partial y}\frac{\partial v}{\partial y} + \frac{\partial^2 u}{\partial y \partial t} - u\frac{\partial^2 v}{\partial x^2} - \frac{\partial u}{\partial x}\frac{\partial v}{\partial y} - \frac{\partial^2 v}{\partial x \partial t}$$

$$= f\left(\frac{\partial v}{\partial y} + \frac{\partial u}{\partial x}\right) + v\frac{\partial f}{\partial y}$$

or

$$u\frac{\partial(\partial u/\partial y - \partial v/\partial x)}{\partial x} + v\frac{\partial(\partial u/\partial y - \partial v/\partial x)}{\partial y} + \frac{\partial u}{\partial x}\left(\frac{\partial u}{\partial y} - \frac{\partial v}{\partial x}\right)$$

$$+ \frac{\partial v}{\partial y}\left(\frac{\partial u}{\partial y} - \frac{\partial v}{\partial x}\right) + \frac{\partial^2 u}{\partial y \partial t} - \frac{\partial^2 v}{\partial x \partial t} = f\left(\frac{\partial u}{\partial x} + \frac{\partial v}{\partial y}\right) + v\frac{\partial f}{\partial y}$$

and

$$-u\frac{\partial \zeta}{\partial x} - v\frac{\partial \zeta}{\partial y} - \zeta\left(\frac{\partial u}{\partial x} + \frac{\partial v}{\partial y}\right) - \frac{\partial \zeta}{\partial t} = f\left(\frac{\partial u}{\partial x} + \frac{\partial v}{\partial y}\right) + \frac{df}{dt}$$

But

$$\frac{d\zeta}{dt} = \frac{\partial \zeta}{\partial t} + u\frac{\partial \zeta}{\partial x} + v\frac{\partial \zeta}{\partial y}$$

so we have

$$\frac{d\zeta}{dt} + \frac{df}{dt} = -(\zeta + f)\left(\frac{\partial u}{\partial x} + \frac{\partial v}{\partial y}\right)$$

or

$$\frac{d(\zeta + f)}{dt} = -(\zeta + f)\,\text{div}\,\mathbf{V} \tag{11.10}$$

In the above derivation we have started from the simplified form of the equations of horizontal motion and neglected spatial variations in the vertical velocity and also variations in the density. Including such variations gives rise to what are called the tilting or twisting terms and the solenoidal terms, respectively. These terms are usually small for synoptic-scale motions.

We see that in deriving equation (11.10) we have eliminated the pressure field altogether. The motion must, of course, be driven by some pressure gradient forcing, but this forcing is outside the terms of reference of the equation. We can then study equation (11.10) as it stands. We note specifically that it relates the rate of change of absolute vorticity to divergence.

The far-reaching practical importance of (11.10) is that we have a means of computing divergence from rates of change of vorticity. Now vorticity is wholly different from divergence in that it can be estimated qualitatively or measured quantitatively from a synoptic chart. This is something that cannot be done with divergence since the only information we can extract from the synoptic chart must use the geostrophic assumption. But we can use the geostrophic assumption to assess the vorticity. Granted, this is an estimate of the geostrophic vorticity, but it does not matter, since the numerical magnitude of vorticity is much greater than that of divergence. We can see this by just looking at weather maps. Anticyclones, cyclones or depressions and wave-like patterns in the free atmosphere all have vorticity which can be immediately recognized.

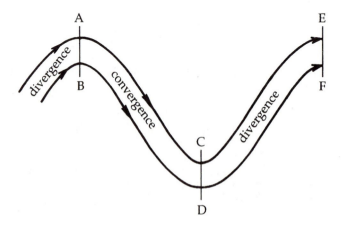

FIGURE 11.7 Convergence and divergence due to curvature in a northern-hemisphere wave pattern.

11.7 A SIMPLE WAVE PATTERN

Figure 11.7 shows a simple wave pattern. Patterns similar to this are a regular feature of upper air charts, for instance at 500 mb (hPa). We will apply the results of the vorticity theorem to this pattern. We will first make some approximations:

$$\frac{d(\zeta + f)}{dt} = \frac{d\zeta}{dt} + \frac{df}{dt} \qquad (11.11)$$

f only varies along the y axis so

$$\frac{df}{dt} = v\frac{\partial f}{\partial y} = \frac{2\omega v \cos \phi}{a} \qquad (11.12)$$

Observations show that in most cases df/dt is small compared with $d\zeta/dt$. Also the relative vorticity is usually numerically smaller than f, and ζ_A is nearly always cyclonic, except occasionally near the equator, or in what might be called singularities in high-intensity jet streams where strong anticyclonic vorticity is generated over a synoptically small spatial volume. We can disregard such anomalies. Equation (11.10) then becomes

$$\frac{d\zeta}{dt} = -f \operatorname{div} \mathbf{V} \qquad (11.13)$$

We will apply equation (11.13) to Fig. 11.7, which is applicable to the northern hemisphere. The isobars are assumed to be equidistant everywhere and the flow is assumed quasi-geostrophic. We also constrain the shape of the pattern so that it retains the same shape in space and time. We will let all the vorticity in the pattern be due to the curvature of the pattern. Let us consider the segment between AB and CD. At the crest of the wave the vorticity is anticyclonic or negative. At the trough it is cyclonic or positive. Clearly, vorticity increases between AB and CD. From (11.13) convergence (negative divergence) must occur in that segment of the wave. If we now consider the segment between CD and EF we note that vorticity is cyclonic or positive at the trough while at the next crest it is again anticyclonic or negative. Vorticity decreases between trough and ridge. We deduce, therefore, from (11.13) that divergence must occur in this segment since the term on the right hand side is now positive. What we see on weather charts does support the foregoing theoretical conclusions. If we look at a long-wave pattern on an upper air chart, say at 500 mb (hPa), it is often noted that a surface anticyclone lies directly downstream between the 500 mb ridge and trough and similarly low-pressure centres at the surface often lie downstream between the 500 mb trough and ridge. Wave-like patterns of this kind are often called long waves.

If we now let Fig. 11.7 apply to the southern hemisphere the crest of the wave at AB becomes the trough and the trough at CD becomes the ridge, and the labels of convergence and divergence in Fig. 11.7 must be interchanged.

It is interesting that the same general conclusions about the location of convergence and divergence in wave patterns may be deduced quite simply from the gradient wind equation. Thus we may write equation (8.5) as

$$f V_{gr} \mp \frac{V_{gr}^2}{R} = f V_g$$

where V_{gr} and V_g refer to the gradient and geostrophic winds respectively, and the negative sign refers to the northern anticyclonic and southern cyclonic cases. Then

$$V_g = V_{gr} \mp \frac{V_{gr}^2}{fR}$$

Thus for the anticyclonic pattern $V_g < V_{gr}$, and for the cyclonic pattern $V_g > V_{gr}$ in both hemispheres. If these relationships are applied to the segments of the wave in Fig. 11.7 we find that the gradient wind speed across AB is greater than across CD. Air will then accumulate within the segment between the ridge and the trough, that is convergence will occur. Similarly the wind is less across CD than across EF leading to a depletion of air between the trough and the ridge, that is divergence.

The observed effect of curvature vorticity is shown as a translation of the pattern from west to east; that is, if a ridge is approaching from the west, then the barometer at the surface will record rising pressure. Conversely, if a trough approaches from the west the barometer at the surface will record falling pressure. If the pattern is constrained to conserve its shape, translation of the wave must occur to explain the response of the barometer. The effect of curvature vorticity in the translation of the wave pattern from west to east in a general westerly current may also be regarded in a physical sense as due to the advection of the relative vorticity by the wind. This will be discussed in a later chapter.

11.8 SHEAR VORTICITY IN A JET STREAM PATTERN

In Fig. 11.7 we assumed that all the vorticity was due to the curvature of the quasi-geostrophic wind. We will now assume a pattern which possesses shear vorticity but no curvature vorticity. Such a pattern may be formed in jet streams which occur at high levels, say between 200 and 350 mb. Such a pattern is shown in schematic form for the northern hemisphere in Fig. 11.8. In the region on the

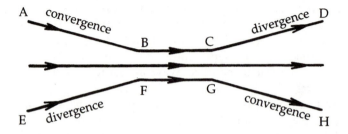

FIGURE 11.8 Convergence and divergence due to shear in a northern-hemisphere jet-type pattern.

polar side of the jet, cyclonic shear increases along AB and decreases along CD. Conversely, in the region equatorward of the jet, anticyclonic shear increases along EF and decreases along GH. Equation (11.13) thus predicts that convergence may be expected to the left of the entrance to the jet and to the right of the exit in the northern hemisphere. Likewise, divergence may be expected to the right of the entrance and to the left of the exit to the jet in that hemisphere.

Figure 11.8 may also represent a jet pattern in the southern hemisphere. In this case the labels for divergence and convergence must be interchanged as in Fig. 11.7. This is because the sign of the shear vorticity is inter-changed.

In actual practice the vorticity distribution around a pattern is due partly to the curvature effect and partly to the shear effect. In some regions of the pattern the two contributions are of opposite sign and tend to cancel each other out. In other regions they reinforce each other.

The convergence and divergence which arise from the vorticity distribution around a given pressure or contour pattern refer to the level or within a layer in which the given pattern exists. At other levels the pattern and consequently the vorticity distribution and resulting convergence and divergence may be quite different. The convergence and divergence calculated in any given level or thin layer may not be very closely related with the rise or fall in surface pressure. For this to occur it would be necessary to calculate and integrate the convergence or divergence for every level from the surface to the outer limits of the atmosphere where pressure was still appreciable. However, the results for a specific level may be significant in estimating development of the surface pressure field if a level is chosen at which ageostrophic motion is a maximum.

The theory of long waves and of their translation will be examined in more detail in the next chapter. This theory is generally applicable to westerly currents. In the troposphere the wind motion is generally westerly except in the tropics, and in rare blocking occasions when large anticyclones extend their structure upwards to 500 or 300 mb levels. Westerly flow arises because of the rotation of the earth and because of the temperature difference beween the poles and the equator, that is the meridional profile of temperature. Long waves rarely occur in an easterly current in temperate latitudes.

11.9 CONSTANT ABSOLUTE VORTICITY TRAJECTORIES

We may construct trajectories of parcels of air which are constrained to conserve their absolute vorticity so that $d(\zeta + f)/dt = 0$. Let us rewrite equation (10.2) omitting a pressure gradient forcing:

$$\ddot{u} + f^2 u = 0$$

The solution is

$$u = u_0 \cos ft + v_0 \sin ft$$
$$v = v_0 \cos ft - u_0 \sin ft$$

(11.14)

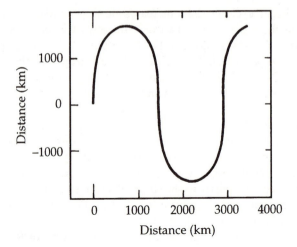

FIGURE 11.9 Constant absolute velocity trajectory of air parcel starting from the equator with a southerly velocity of $5\,\text{m}\,\text{s}^{-1}$.

where u_0, v_0 are some initial velocities and

$$x = \frac{u_0}{f}\sin ft + \frac{v_0}{f}(1 - \cos ft)$$

$$y = \frac{v_0}{f}\sin ft - \frac{u_0}{f}(1 - \cos ft)$$

(11.15)

Equations (11.14) and (11.15) above express the velocities and trajectories of air parcels which are projected at an initial velocity u_0, v_0 from some initial origin of coordinates. The tracks are then constant absolute vorticity trajectories. We may arbitrarily assign such initial velocities and compute the tracks. If f is held constant the trajectories are inertial circles. If f is allowed to vary the trajectories are more accurate, particularly if they spend any time in equatorial latitudes. These trajectories conserve their absolute vorticity and so obey the constraint

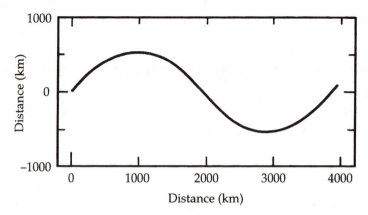

FIGURE 11.10 As for Fig. 11.9 but starting with a southwesterly of $10\,\text{m}\,\text{s}^{-1}$.

$d(\zeta + f)/dt = 0$. These air parcels start from the equator and so they have zero total vorticity at all times. In Fig. 11.9 the initial velocity is $5\,\mathrm{m\,s^{-1}}$ from the south. In Fig. 11.10 the initial velocity is $u_0 = 5\,\mathrm{m\,s^{-1}}$, $v_0 = 5\,\mathrm{m\,s^{-1}}$ giving a southwest wind of $7.07\,\mathrm{m\,s^{-1}}$.

Now suppose we replace u_0, v_0 in equations (11.14) and (11.15) above by u_g, v_g. We may then presume that the absolute vorticity trajectories we calculate are trainlines already laid down by nature in determining the planetary flow pattern.

In the next chapter we will investigate in more physical and mathematical detail the properties of wave patterns in the free atmosphere under the forcing for which the air is assumed to follow constant absolute vorticity trajectories.

11.10 PROBLEMS

1. Under what conditions does the advection of relative vorticity equal the product of the absolute vorticity and the horizontal divergence?
2. What is the vorticity of a disc rotating at 10 revolutions per second?
3. The geostrophic wind is westerly and changes from $25\,\mathrm{m\,s^{-1}}$ at $40°\mathrm{N}$ to $5\,\mathrm{m\,s^{-1}}$ at $50°\mathrm{N}$. What is the magnitude and sign of the relative vorticity if the rate of change of the wind velocity is constant? What would be the radius of an isobaric system having the same relative vorticity if the gradient wind velocity was $10\,\mathrm{m\,s^{-1}}$, assuming no shear?
4. Derive an expression for the relative vorticity of the gradient wind from the gradient wind equation.
5. What is the vorticity of flow represented by the stream function?
6. What is the relative vorticity of the outer structure of the tornado in problem 5 of Chapter 8 if $n = 1$?
7. Write the divergence vorticity equation in terms of the stream function.

THE LONG-WAVE EQUATIONS

The derivation of the long-wave equations is attributed to Carl Rossby, an outstanding meteorologist of the twentieth century. They afforded a breakthrough in thinking of meteorology after World War II. Such waves are often called Rossby waves. The theory developed here relies on the theorem of the conservation of absolute vorticity.

The breakthrough in thinking came largely as a result of the invention of the radiosonde instrument which enabled temperatures and pressures to be measured in the upper atmosphere, and transmitted by radio to a ground-based receiver. From the equation of state we have already seen that if two of the three variables p, T, ρ are known, we can calculate the third variable. Knowing the pressure and temperature at different heights we can calculate the density and also the pressure gradients. But more important, as we shall see in the next chapter, we can use the thickness equation to calculate the heights of the pressure surfaces. Either way it becomes possible to construct upper air charts of the pressure field, or of the height field, and use the geostrophic assumption to calculate the wind field. It was immediately noted that the quasi-geostrophic flow pattern along the isopleths of pressure or height possessed a wave-like structure and that these waves meandered around the globe.

12.2 EFFECTS OF CURVATURE AND LATITUDE VORTICITY ON WAVE TRANSLATION

We have already seen in Fig. 11.7 that curvature vorticity produces convergence in that part of a wave pattern downwind of the ridge and upwind of the trough, and divergence in that part of the pattern downwind of the trough and upwind of the ridge. We were able to show this effect first from the vorticity equation and secondly from the gradient wind equation. We also said that the advection of the curvature or relative vorticity in the pattern, the shape of which remains constant in time and space, caused the pattern to translate from west to east. We may now use the simple geostrophic equation to show that the earth vorticity produces the opposite effect. We assume a northern-hemisphere long-wave pattern (Fig. 12.1) where the isobars are spaced at an equal distance from one another. Then, at the ridge line AB

$$v_{g_1} = \frac{1}{(\rho f_1)} \frac{\delta p}{\delta n}$$

and at the trough line CD

$$v_{g_2} = \frac{1}{(\rho f_2)} \frac{\delta p}{\delta n}$$

where v_{g_1}, v_{g_2}, are the respective geostrophic velocities and δn is the distance between the isobars. Dividing the first of the above relations by the second we have

$$\frac{v_{g_1}}{v_{g_2}} = \frac{f_2}{f_1} = \frac{\sin \phi_2}{\sin \phi_1}$$

$$v_{g_1} = v_{g_2} \frac{\sin \phi_2}{\sin \phi_1}$$

(12.1)

but $\sin \phi_2 < \sin \phi_1$ and so $v_{g_1} < v_{g_2}$.

Thus, in this case there is divergence between AB and CD and convergence between CD and EF. We see that this is just the opposite of the effect shown in Fig. 11.7. The effect shown in Fig. 12.1 is due to the latitude vorticity, that is to the rotation of the earth. The distribution of convergence and divergence in Fig. 12.1 may also be determined from equation (11.13), as vorticity now increases upstream from AB and decreases between AB and CD. The pattern will be translated from east to west as the advection of f, the planetary vorticity, by the wind is in the opposite direction to the advection of the curvature vorticity. In this case the pattern is advected against the wind field. The net result of these two opposing effects may be as illustrated in Fig. 12.2. Figure 12.2(a) shows a

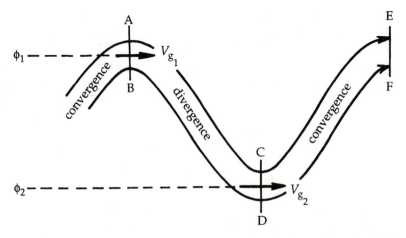

FIGURE 12.1 The latitude effect in a long-wave pattern in the northern hemisphere.

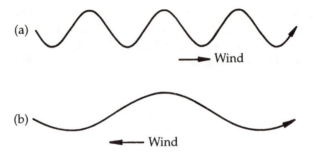

FIGURE 12.2 (a) Short wave translated in the direction of the wind; (b) long wave translated against the wind.

short-wave pattern. In such a pattern the effect of changes in the curvature vorticity exceeds the effect of changes in the latitude vorticity. The pattern will move from west to east. The opposite happens in Fig. 12.2(b). There, the effect of changes in the curvature vorticity is small compared with changes in the latitude vorticity. The pattern will therefore move from east to west. The general result is that short-wave patterns move from west to east. On the other hand, long-wave patterns move from east to west, upstream against the wind. Such movement is sometimes called retrogressive. These predicted motions are observed to be the case when we follow the movement of such waves on upper air weather charts.

These principles form the basis of the Rossby long-wave equations. Their identification provided an extremely important breakthrough in the under-standing of the way in which the atmosphere behaves. The theoretical and practical applications of these ideas turned out to be of the utmost significance in modern dynamic meteorology. A more exact mathematical derivation of the general ideas discussed above will now be given.

12.3 THE ROSSBY LONG-WAVE EQUATION

The development of the Rossby long-wave equation relies on the principle of the conservation of absolute vorticity, which was introduced in the previous chapter. Absolute vorticity may be conserved in nature as can such quantities as mass, heat and momentum. To conserve absolute vorticity we will write the relation $d(\zeta + f)/dt = 0$. This was done in the preceding chapter when we computed constant absolute vorticity trajectories. In order for absolute vorticity to be conserved the divergence must be zero from the vorticity equation. We must therefore apply the theorem to some level where the flow is non-divergent. We saw in Fig. 9.4 that there is usually some level at which the divergence profile changes sign. This is often found to be at about 600 mb (hPa). This level may be considered as a level equivalent to or representative of a simplified mean atmosphere.

12.4 THE LONG-WAVE THEORY

Consider a long-wave pattern at the level of non-divergence (9.3) which is normally to be found at about 600 mb. Then

$$\frac{d(\zeta + f)}{dt} = 0$$

and

$$\frac{\partial \zeta}{\partial t} + u \frac{\partial \zeta}{\partial x} + v \frac{\partial \zeta}{\partial y} + v \frac{\partial f}{\partial y} = 0$$

It follows that the total vorticity must be conserved at this level. We may transform the term

$$\frac{\partial f}{\partial y} = \frac{2\Omega \cos \phi}{R} = \beta$$

where R is the radius of the earth, so that

$$v \frac{\partial f}{\partial y} = \beta v \tag{12.2}$$

β expresses the instantaneous rate of change of the earth's vorticity at a given point in the flow.

Equation (12.2) now becomes

$$\frac{\partial \zeta}{\partial t} + u \frac{\partial \zeta}{\partial x} + v \frac{\partial \zeta}{\partial y} + v\beta = 0 \tag{12.3}$$

It is assumed that there is a zonal current of uniform constant velocity U upon which is superimposed a perturbation with the velocity components u' and v'. The total velocities, which are assumed independent of y, are then

$$u = U + u'$$
$$v = v' \tag{12.4}$$

Now the mean zonal velocity is independent of x and y so that from (11.7) and (12.3)

$$\zeta = \frac{\partial v'}{\partial x} - \frac{\partial u'}{\partial y} = \zeta'$$

It follows from (12.3) and (12.4) that

$$\frac{\partial \zeta}{\partial t} + (U + u') \frac{\partial \zeta}{\partial x} + v' \frac{\partial \zeta}{\partial y} + v'\beta = 0 \tag{12.5}$$

The term $u'(\partial \zeta / \partial y) = 0$, since the vorticity is independent of y.

We have left

$$\frac{\partial \zeta}{\partial t} + U \frac{\partial \zeta}{\partial x} + v'\beta = 0 \tag{12.6}$$

Now, if the perturbations are travelling eastward without change of shape it follows that

$$c\frac{\partial \zeta}{\partial x} = -\frac{\partial \zeta}{\partial t} \tag{12.7}$$

where c is equal to the eastward velocity of the pattern.

Substituting (12.6) in (12.7)

$$U\frac{\partial \zeta}{\partial x} - c\frac{\partial \zeta}{\partial x} + v'\beta = 0 \tag{12.8}$$

It was assumed that the perturbations are independent of y. Then

$$\frac{\partial u'}{\partial y} = 0$$

and

$$\zeta = \frac{\partial v'}{\partial x}$$
$$(U - c)\frac{\partial^2 v'}{\partial x^2} + v'\beta = 0 \tag{12.9}$$

This is a differential equation the solution of which may be given by an expression of type

$$v' = v'_0 \sin\frac{2\pi}{L}(x - ct) \tag{12.10}$$

where v'_0 is a constant.

If (12.9) is differentiated twice with respect to x, then

$$\frac{\partial v'}{\partial x} = v'_0 \frac{2\pi}{L} \cos\frac{2\pi}{L}(x - ct)$$
$$\frac{\partial^2 v'}{\partial x^2} = -v'_0 \frac{4\pi^2}{L^2} \sin\frac{2\pi}{L}(x - ct)$$

Substituting (12.9) in (12.8)

$$-(U - c)v'_0 \frac{4\pi^2}{L^2} \sin\frac{2\pi}{L}(x - ct) = -v'_0 \beta \sin\frac{2\pi}{L}(x - ct)$$

$$U - C = \frac{\beta L^2}{4\pi^2}$$

or

$$C = U - \frac{\beta L^2}{4\pi^2} \tag{12.11}$$

where L is the wavelength of the sinusoidal disturbance.

12.5 THE STATIONARY WAVELENGTH

The formula (12.11) expresses the velocity of the long wave in terms of the mean zonal speed and the wavelength. If the wave is stationary $c = 0$ and (12.11) becomes

$$U = \frac{\beta L_s^2}{4\pi^2} \tag{12.12}$$

where L_s denotes the stationary wavelength

$$L_s = 2\pi \sqrt{\frac{U}{\beta}} \tag{12.13}$$

Substituting for U from (12.13) in (12.11) we have

$$c = \frac{\beta L_s^2}{4\pi^2} - \frac{\beta L^2}{4\pi^2} \tag{12.14}$$

It is clear from (12.14) that if

$$L_s > L \quad \text{then} \quad c > 0$$
$$L_s < L \quad \text{then} \quad c < 0$$

Thus, if the wavelength is less than the stationary wavelength the wave velocity is positive, that is from west to east. If, on the other hand, the wavelength is longer than the stationary wavelength for the selected latitude and mean wind speed, the velocity is negative, that is from east to west (retrogressive).

It is seen from equation (12.13) that the stationary wavelength increases with zonal wind speed and with latitude. Thus in high latitudes the wave velocity will be greater for a given wind speed than in low latitudes for a given wavelength. In temperate latitudes where strong jet streams occur the waves will move fast from west to east since both the mean zonal wind speed and the latitude are larger. In latitudes nearer the tropics they will move more slowly from west to east, or remain almost stationary, or even move slowly from east to west if the zonal wind speed is small and the actual wavelength is longer than the stationary wavelength.

Although the Rossby long-wave theory as presented here has been simplified, and some assumptions have been made, the results do give an excellent insight into the way in which long waves in the troposphere behave. Since the crests of the long waves in the 500 mb (hPa) chart are associated with anticyclones at the surface, and the troughs are associated with low-pressure centres at the surface, the computed speed of the wave will tell us its anticipated position 12 to 24 hours ahead.

To summarize, why do shortwaves travel towards the east and why does the eastward velocity increase as the wavelength decreases? The answer is that the combined velocity of the wave and the wind through the geostrophic control pattern must be a constant absolute vorticity trajectory. Absolute vorticity must be conserved as that was the foundation stone of the theoretical development. Why does a wave which is longer than the stationary wavelength travel towards the west? The same answer applies. In order to maintain a constant absolute vorticity trajectory the long wave must move in the opposite direction to that of the air current through the geostrophic control pattern. In other words, $d(\zeta + f)/dt = 0$.

12.6 ABSOLUTE VORTICITY OF LAYER OF CONSTANT MASS

In the discussion so far we have derived the term $d(\zeta + f)/dt = 0$ which expresses the theorem of the conservation of absolute vorticity. The formula has been derived for some constant level of height or of pressure. We may, however, derive a more general expression which is less restrictive and therefore more useful as a tool in the analysis and prediction of movement of long waves.

We return to the vorticity equation (11.10)

$$\frac{d(\zeta + f)}{dt} = -(\zeta + f)\,\mathrm{div}\,\mathbf{V}$$

We have already shown in equation (10.12) that horizontal divergence may be expressed as

$$\mathrm{div}\,\mathbf{V}_{\mathrm{H}} = \frac{1}{A}\frac{\mathrm{d}A}{\mathrm{d}t}$$

where A is a given area. Hence,

$$\frac{1}{(\zeta + f)}\frac{\mathrm{d}(\zeta + f)}{\mathrm{d}t} = -\frac{1}{A}\cdot\frac{\mathrm{d}A}{\mathrm{d}t}$$

$$\log(\zeta + f) = -\log A + \mathrm{constant}\quad \log A(\zeta + f) = \mathrm{constant}$$

$$A(\zeta + f) = \mathrm{constant}$$

and $A = M/\rho\,\mathrm{d}z = Mg/\mathrm{d}p$. Therefore, since mass is conserved,

$$\frac{(\zeta + f)}{\mathrm{d}p} = \mathrm{constant} \tag{12.15}$$

and hence

$$\frac{\mathrm{d}(\zeta + f)}{\mathrm{d}p} = 0 \tag{12.16}$$

Equation (12.16) is a more explicit expression for the conservation of absolute vorticity than (11.10) as it is not confined to a given height or pressure level. It embodies a given mass or thickness of air measured in millibars or hectopascals. We shall see in the next chapter that all upper air charts are analysed at selected pressure levels and that numerical models are concerned with layers bounded by two pressure surfaces. In approximate studies these layers may be bounded by the 1000 and 500 mb (hPa) layers, or by the 500 and 300 mb (hPa) layers, and so forth. In more accurate models, say 10 layers, or even 20 layer models, the thicknesses are 100 mb or 50 mb, respectively. Thus, the vorticity equation in this form enables constant vorticity trajectories and, thus, Rossby long waves to be represented for a given layer or slice of the atmosphere. The assumption of non-divergence is automatically included in the method of representing the atmosphere as made up of layers or slices of constant mass; that is, the slices are enclosed by fixed upper and lower pressure levels.

12.7 POTENTIAL VORTICITY

For an incompressible fluid such as the ocean $(\zeta + f)/dp$ may be defined as potential vorticity. In oceanographic work dp is usually replaced by D, the depth, since $dA/A = dh/h = dD/D$ in the development of (12.16) in the preceding section.

The concept of potential vorticity applied to the atmosphere requires a slightly different treatment since the atmosphere is not incompressible. We must now consider a slice of air bounded by constant isentropic surfaces, that is surfaces of constant potential temperature, and assume that the air is subject to adiabatic motion.

From the preceding section we have found that $A(\zeta + f) = $ constant but

$$A = \frac{Mg}{\delta p} = \frac{\delta\theta}{\delta p} \times \frac{Mg}{\delta\theta}$$

$Mg/\delta\theta$ is constant since the difference between the two isentropic surfaces is constant. Therefore $A = $ constant $\times \delta\theta/\delta p$ and

$$(\zeta + f)\frac{\partial\theta}{\partial p} = \text{constant} \qquad (12.17)$$

and

$$\frac{d}{dt}\left(\frac{(\zeta + f)\partial\theta}{\partial p}\right) = 0 \qquad (12.18)$$

Equation (12.18) expresses the conservation of potential vorticity in an atmosphere in which the motion is adiabatic. It is a measure of the ratio of the absolute vorticity to the depth of the vortex. It is an important concept because it tells us how the vorticity changes as the thickness of the vortex between two isentropic surfaces changes. This parameter is referred to frequently in modern dynamic meteorology.

12.8 PROBLEMS

1. What is the wavelength of a stationary Rossby wave at 45°N if the mean wind speed is westerly at $10\,\mathrm{m\,s^{-1}}$?
2. What is the velocity of a Rossby wave at 45°N if the wavelength is 5000 km and the mean zonal westerly wind is $25\,\mathrm{m\,s^{-1}}$?
3. In Fig. 11.10 the wavelength of the constant absolute vorticity trajectory was about 3000 km. What would be the wave speed? If the wavelength of the forcing geostrophic pattern was half of the above wavelength what would be its velocity?

13

THE UPPER AIR SYNOPTIC CHART

13.1 INTRODUCTION

In the preceding chapters we have confined our treatment of the flow of the atmosphere to motion referred to a horizontal plane. Although this concept works well for the surface weather map, it is difficult to apply it to upper air charts which also involve the dimension of height. As we shall now discuss the motion of the atmosphere as a whole, that is in three dimensions, we must include this third dimension into our deliberations. In order to obtain geostrophic wind speeds we need to know the pressure gradients at different levels. The radiosonde instrument measures pressure and temperature, but not height directly. There is also the difficulty of computing the density at different heights. It would involve tedious calculations to obtain the pressure and density of selected reference height levels. These difficulties were realized internationally at the end of World War II and a much better scheme was devised, not only for the plotting and analysis of upper air charts, but also for their interpretation.

13.2 PRESSURE AS A VERTICAL COORDINATE

Consider a point P in the isobaric surface at a height z above m.s.l. Next let Q be a nearby point in the same surface at a height $z + \mathrm{d}z$ above m.s.l. and let us suppose that it is situated so that PQ represents the direction of steepest slope of the isobaric surface at P (Fig. 13.1). Then, if Q′ lies directly beneath Q at a height z above m.s.l. and if $\mathrm{d}n$ represents the infinitesimal distance PQ′, we have that the pressure change on going from P to Q via Q′ is

$$\mathrm{d}p = \frac{\partial p}{\partial n}\,\mathrm{d}n + \frac{\partial p}{\partial z}\mathrm{d}z = 0 \qquad (13.1)$$

since we are moving on a surface of constant pressure. From the hydrostatic

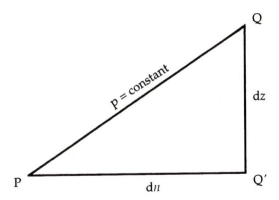

FIGURE 13.1 Schematic of isobaric surface sloping in space.

equation and the fact that $dz = (\partial z/\partial n)\, dn$ we see from (13.1) that

$$\frac{\partial p}{\partial n} - g\rho\,\frac{\partial z}{\partial n} = 0$$

or

$$-\frac{1}{\rho}\frac{\partial p}{\partial n} = -g\frac{\partial z}{\partial n} \tag{13.2}$$

We may therefore rewrite the geostrophic wind as

$$u_g = -g\frac{\partial z}{\partial y} \tag{13.3}$$

$$v_g = g\frac{\partial z}{\partial x} \tag{13.4}$$

This is a most important and useful advance. We have reduced the number of variables needed to calculate the geostrophic wind from three to two as we have eliminated density from our equation. We now only need to know the gradient of the height of the pressure surface since g is a constant. In order to take advantage of this simpler method we must transform our means of constructing our upper air charts from maps of the pressure at specified heights to maps showing the heights of specified isobaric surfaces, that is surfaces of constant pressure, above m.s.l. Isopleths of constant height on a map are called contours. We shall therefore refer to contours on synoptic charts of the troposphere in the same way as contours on a geographical map of a country or continent of the world. We shall see that this different representation opens up many new avenues of exploration in the understanding of the dynamics of the weather.

13.3 THE THERMAL WIND

A new parameter of great importance in atmospheric dynamics may now be introduced. It is a measure of the vertical variation of the geostrophic wind.

We will consider the geostrophic wind at two levels. Let u_{g_1}, v_{g_1} be the geostrophic wind at some higher level p_1, z_1 and u_{g_0}, v_{g_0} be the geostrophic wind at some lower level p_0, z_0. Then

$$u_{g_1} - u_{g_0} = \frac{-g}{f}\left(\frac{\partial z_1}{\partial y} - \frac{\partial z_0}{\partial y}\right) \tag{13.5}$$

and

$$v_{g_1} - v_{g_0} = \frac{g}{f}\left(\frac{\partial z_1}{\partial x} - \frac{\partial z_0}{\partial x}\right) \tag{13.6}$$

or

$$\delta u_g = \frac{-g}{f}\frac{\partial(\delta z)}{\partial y}, \quad \delta v_g = \frac{g}{f}\frac{\partial(\delta z)}{\partial x} \tag{13.7}$$

Table 13.1 may be used to convert a synoptic chart of the m.s.l. pressure distribution into a chart of the contour pattern of the 1000 mb isobaric surface. The chart then represents the topography of the 1000 mb chart in the same sense as a geography map may represent the topography of the underlying surface by height contours. We may continue with this method and use the thickness equation to compute the thicknesses of different layers of the troposphere. We may then add the different thicknesses together and obtain the total heights of the different standard pressure levels above m.s.l. Contours of the heights of the pressure surfaces may then be constructed so that we have topographies of the heights of the standard pressure surfaces above m.s.l. There is one problem to overcome here. It is to find the mean temperature of a layer.

13.4 THE THICKNESS OF A STANDARD ISOBARIC LAYER

Suppose the heavy zig-zag curve in Fig. 13.2 represents an actual virtual temperature distribution starting from T_1^* at pressure p_1 and finishing at T_2^* on the p_2 isobar. If we were to replace this distribution of temperature by an equivalent isothermal one, which can be done by finding the isotherm AQB for which the shaded areas AQT_1^* and BQT_2^* are equal, then the thickness of the original layer and that of the equivalent isothermal layer will be the same, since the same area A lies to the left of each temperature curve.

Since the thickness of the equivalent isothermal layer can be calculated from the thickness equation we can construct a thickness scale parallel to the isobars as shown in the figure. It is often hard to estimate the temperature of an equivalent isothermal layer accurately, when large areas such as those in Fig. 13.2 have to be balanced. In practice, since the thickness scale is placed

TABLE 13.1 Transformation of an m.s.l. synoptic pressure pattern to a topography of the 1000 mb surface using the conversion 8 mb = 60 m

Sea-level pressure	960	968	976	984	992	1000	1008	1016	1024	1032	1040
Height of 1000 mb surface (m)	−300	−240	−180	−120	−60	0	60	120	180	240	300

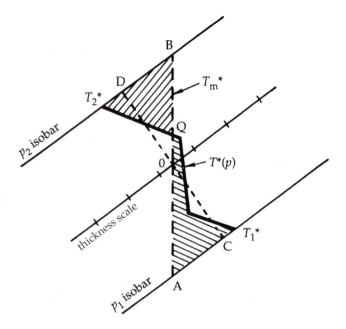

Figure 13.2 The thickness of a standard isobaric layer.

geometrically midway between the isobars, the intersection AOB can be replaced by a line closer to the actual temperature curve, such as COD, and because the area $\triangle AOC = \triangle BOD$, the intersection of COD with the scale will still give the correct thickness.

The graphical computation of thicknesses can be done on any tephigram, Emagram or skew-log p diagram upon which the virtual temperature structure has been plotted.

Thus, the thermodynamically determined thicknesses may be added cumulatively to construct the topography of various standard isobaric surfaces. In this way we may draw upper air charts of the 850, 700, 500, 300, 200 and 100 mb surfaces. In the analysis of routine daily charts the 1000–500 thickness is important as it represents the lower half of the mass of the atmosphere.

13.5 DIFFERENTIAL ANALYSIS OF THE UPPER AIR SYNOPTIC CHART

The basic upper air synoptic chart is a contour chart of the 500 mb isobaric surface. As we have seen in Section 13.3 a surface map of the m.s.l. pressure distribution may easily be transformed into a contour chart of the 1000 mb height field. If we compute the 1000–500 thickness as described in the preceding section we may add the two fields together to get the 500 mb contour map. Figure 13.3 is applicable to both hemispheres. It shows how the summation may be carried out graphically. It is seen that a third set of lines can be drawn through the intersection points of the thickness lines and the 1000 mb contours. This

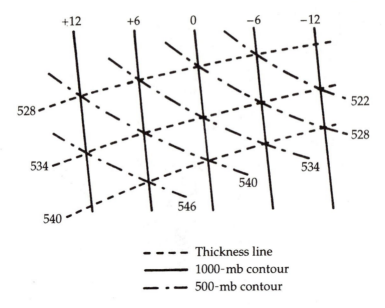

FIGURE 13.3 The gridding technique (heights in dm).

technique of differential analysis has been called 'gridding'. We will now blow up one of the cells in Fig. 13.3 as shown in Fig. 13.4. The northern-hemisphere winds are shown by arrows to signify that they are vectors, that is they possess both speed and direction. In the chosen cell the surface geostrophic wind is denoted by V_0. It blows from the north. The thermal wind which blows along the thickness lines is denoted by V'. It is mainly westerly indicating that there is colder air to

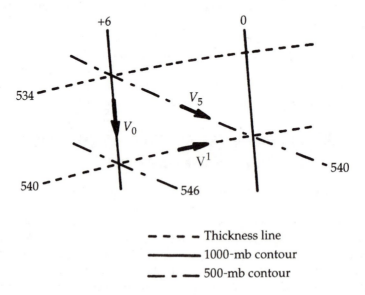

FIGURE 13.4 The thermal wind for the northern-hemisphere case.

the north. The thermal gradient is therefore directed normal to the thickness lines. If we add the two vector winds

$$\mathbf{V}_0 + \mathbf{V}' = \mathbf{V}_5 \qquad (13.8)$$

and

$$\mathbf{V}' = \mathbf{V}_5 - \mathbf{V}_0 \qquad (13.9)$$

we see then that the thermal wind is the vector difference between the geostrophic winds at 500 mb and the surface. This is the theoretical result we obtained when deriving the mathematical form of the thermal wind equation (13.5). Thus, it is emphasized that the thermal wind behaves in the same way as the geostrophic wind in that it blows along the thickness lines. In the northern hemisphere low temperature is to the left of the flow and warm temperature is to the right of the flow. In the southern hemisphere the reverse is true. The velocity of the thermal wind is inversely proportional to the perpendicular distance between the thickness lines, as the velocity of the geostrophic wind is inversely proportional to the perpendicular distance between the contours.

13.6 BAROTROPIC AND BAROCLINIC STRUCTURE

Barotropic and baroclinic are words which are used to describe two different states of the atmosphere. Simple diagrams based on the construction of upper air charts will bring out the properties of these two kinds of atmospheres.

Concisely, a barotropic atmosphere exists if there is no thermal wind. A baroclinic atmosphere exists if there is a thermal wind. Alternatively, a barotropic atmosphere may be defined as one in which isopleths of density or specific volume are parallel to the isobars. A baroclinic atmosphere may then be defined as one in which isopleths of constant density or specific volume intersect the isobars. The intersection creates solenoids. The latter are exhibited as geometrical areas. Examples of such solenoidal areas are shown in Fig. 13.3. The greater the thickness gradient, the greater is the number of solenoids and the greater the amount of available energy.

If there is no variation of the geostrophic wind with height and therefore no thermal wind, the contour pattern will look the same at all levels as it does at the surface. The surface pressure map will be reproduced at 500 mb. One may compare this structure with a brick wall. No matter where an additional brick is placed within a vertical column of bricks, the net effect of the extra brick will be felt at all levels. The surface pressure underneath the bottom brick will register one brick more while the top of the column will be one brick higher. Such a situation is shown in Fig. 13.5. Figure 13.6 shows a vertical cross-section along the east–west axis of a baroclinic structure. The thickness increases towards the east giving rise to a southerly thermal wind in the northern hemisphere. We see that the actual wind at the left hand or easterly side of the diagram is northerly at the surface and southerly at 500 mb for the northern-hemisphere case. The change in wind between the two levels is the thermal wind. The structure of the atmosphere may therefore be divided into two components, the barotropic and the baroclinic. Any pattern of isobars or contours on the

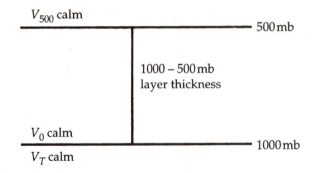

FIGURE 13.5 Vertical cross-section of a thickness layer – no thermal wind.

barotropic component, which is always exhibited by the surface m.s.l. pressure chart, is solely due to the horizontal distribution of mass. On the other hand, any pattern exhibited by the thickness pattern is solely due to the horizontal distribution of the mean temperature of that thickness layer. In any real case the patterns shown by the upper level contours are a combination of the two kinds of structure, barotropic and baroclinic. A considerable barotropic component occurs sometimes in large blocking anticyclones, such as the subtropical anticyclones, and in the tropics. However, the energy is provided by the baroclinic component. In a true barotropic atmosphere the circulation would spin down in a matter of weeks, that is the e-folding time, or the time needed for the wind circulation to decrease to $1/e$ of its original value. This decay in circulation would be caused by surface friction, a subject which we will introduce in the next chapter.

13.7 ADVECTION OF THICKNESS LINES

The construction of an upper level contour chart for some period in the future, say 12, 24 or 36 hours, must depend on the combined behaviour of the two structural components of the atmosphere, the barotropic and the baroclinic.

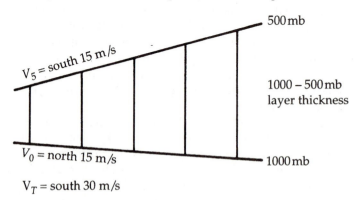

FIGURE 13.6 Vertical cross-section of a thickness layer – strong southerly thermal wind (northern hemisphere case).

Long-wave theory, as derived in the last chapter, may assist in predicting the behaviour of the upper long-wave pattern exhibited by the upper air contours. However, another factor now emerges. It is obvious that the thickness pattern must change with time in a moving atmosphere. The thickness column at a given place is dependent on the mean temperature of the layer. But the whole air column above a fixed point is constantly changing as the wind at different levels is blowing at different speeds and in different directions. The mean temperature or thickness of the air column above must also change as different parts of the column are replaced by air of different temperature. Thus the thickness pattern itself must change and this will change the baroclinic component and, in turn, the actual flow pattern as shown by the contour fields at different levels. The thickness lines must be blown along, or advected, by the wind field caused by the barotropic part of the structure. In other words, the surface wind field advects the thickness pattern. This effect is shown in the solenoidal cell shown in Fig. 13.4.

Advection is well expressed by a simple dot product vector

$$\mathbf{V} \cdot \nabla \bar{T}$$

where $\bar{T}$ is the mean temperature or the value of a given isopleth of thickness. We have met with advection early in this work. For example, the acceleration terms in the equations of motion contain advective terms. Thus,

$$\frac{\mathrm{d}u}{\mathrm{d}t} = \frac{\partial u}{\partial t} + \mathbf{V} \cdot \nabla u \tag{13.10}$$

and

$$\frac{\mathrm{d}v}{\mathrm{d}t} = \frac{\partial v}{\partial t} + \mathbf{V} \cdot \nabla v \tag{13.11}$$

We mentioned in Chapter 11 that vorticity may be advected. Various other quantities may be advected, like heat, water vapour, pollutants, trace gases, etc. Note that a dot product in a vector expression gives a scalar quantity for the product. The dot product also indicates that the magnitude of the advection is the product of the wind and the gradient of the quantity being advected times the cosine of the angle between the directions of the wind vector and the gradient concerned. If the wind is perpendicular to the thickness lines then the advection is $V_0 |\nabla \bar{T}|$. We may therefore formulate the rule that the thickness pattern is advected by the component of the surface wind normal to the thickness line. As we have seen this is the barotropic component of the wind.

An example of northern-hemisphere cold-air advection can be seen in Fig. 13.3. Since the 1000 mb surface is 120 metres above m.s.l. in the eastern (left hand edge of the diagram) and 120 metres below m.s.l. on the western (right hand edge) and the diagram relates to the northern hemisphere, the surface geostrophic wind is northerly and the cold air shown by the smaller thickness values is being advected, or blown southwards.

Thus we do have a means by which we can predict the future shape of the thickness pattern. We simply advect the pattern with the wind at every point. There are, of course, other factors to be borne in mind in advecting thickness patterns. If cold air is advected over a warm sea surface the advection will be less than the formula indicates. Conversely, if cold air is advected over a colder sea surface the thickness lines will tend to move faster than the formula indicates. Similar considerations apply to the advection of warm air over cold and warm surfaces.

13.8 M.S.L. PRESSURE MAPS VERSUS TOPOGRAPHY OF 1000 MB CHARTS

It is pertinent at this stage to comment on the relative merits of the historical and conventional method of drawing isobars and the newer method of transforming the pressure chart to one of the height of the 1000 mb surface. First, surface observations must be made by barometers. M.s.l. pressure readings are therefore available for plotting on a map. Everyone is familiar with weather maps containing isobars. They are published daily in newspapers and are seen on television. Since the number of weather stations reporting conventional surface measurements is far greater than the number of upper air radiosonde reporting stations, it has been found more convenient to retain weather maps of the pressure field. Also, a great deal of historical data obtained from ships' logs over the oceans, as well as from land stations, is being compiled in the investigation of climate change. It would be pointless to convert such a vast amount of data from their original form.

13.9 VORTICITY ON ISOBARIC SURFACES

The vorticity about a vertical axis in a horizontal plane is expressed by

$$\zeta = \frac{\partial v}{\partial x} - \frac{\partial u}{\partial y} \tag{13.12}$$

and the geostrophic wind by

$$u_g = \frac{-g}{f} \frac{\partial z}{\partial y} \tag{13.13}$$

$$v_g = \frac{g}{f} \frac{\partial z}{\partial x} \tag{13.14}$$

Substituting the geostrophic velocities into the expression for vorticity and neglecting the Meridinal component of the Coriolis parameter (i.e. assuming f to be constant), we have

$$\zeta = \frac{g}{f} \left(\frac{\partial^2 z}{\partial x^2} + \frac{\partial^2 z}{\partial y^2} \right) = \frac{g}{f} \nabla^2 h \tag{13.15}$$

where the height notation z has been replaced by h, the height of a grid point on an isobaric surface on a contour chart. Thus, vorticity on an isobaric surface is simply a constant times the Laplacian of the height field. Equation (13.7) can be conveniently transformed into a finite-difference formula.

Consider a grid composed of arms of equal length such as shown in Fig. 13.7. Then,

$$u_g = -\frac{g}{2bf}(h_1 - h_3)$$

$$v_g = \frac{g}{2bf}(h_2 - h_4)$$

(13.16)

where b is the length of the arms; this represents the mean geostrophic wind field covering the grid. Substituting finite differences of velocity as obtained from (13.7) there follows

$$\zeta = \frac{g}{b^2f}[(h_2 - h_5) - (h_5 - h_4) + (h_1 - h_5) - (h_5 - h_3)]$$

$$\zeta = \frac{g}{b^2f}(h_1 + h_2 + h_3 + h_4 - 4h_5)$$

(13.17)

This represents the geostrophic vorticity field within the grid about the point h_5.

The formula (13.17) is a convenient one for calculating vorticity numerically from a contour chart. It is merely necessary to add the heights at the four arms of the grid and subtract four times the value at the centre, and multiply by the correct factor. A scale on the lines of Fig. 13.7 may be constructed so that height values may be interpolated from the contour chart.

13.10 THE VELOCITY POTENTIAL

In Section 10.7 we expressed the wind in terms of a stream function. It is easily seen that the divergence of the stream function component of the wind is zero. If

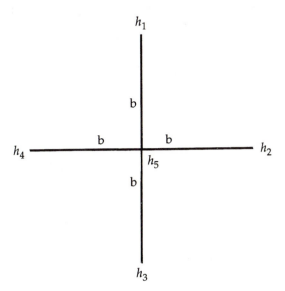

FIGURE 13.7 Height grid for use with contour chart.

the wind field is constructed on a weather map in terms of its stream function, such a wind pattern must be non-divergent. Thus a stream function pattern shows the non-divergent component of the wind.

We may also form an expression which describes the divergent component of the wind. Thus

$$u = \frac{\partial \chi}{\partial x}$$

$$v = \frac{\partial \chi}{\partial y}$$

$$(13.18)$$

where χ is defined as a velocity potential. It is readily seen that the Laplacian of the velocity potential is the divergence. Thus centres of low velocity potential are areas of strong divergence and centres of high velocity potential are areas of strong convergence. This can be visualized if one regards the wind as blowing up and perpendicular to the gradient of the velocity potential isopleths with a speed that is proportional to the gradient. Isopleths of the velocity potential are often constructed for upper levels on synoptic charts in order to delineate regions of divergence and convergence. In Sir Horace Lamb's classical hydrodynamics treatise the signs on the right hand side of (13.10) are negative so that the wind blows down the gradient. This reverses the sign of the divergence/convergence at the high and low centres of the velocity potential pattern. Since divergence is positive and convergence is negative the latter representation might seem more consistent from a meteorological point of view.

13.11 PROBLEMS

1. If the pressure gradient acceleration is 0.001 what is the pressure gradient in mb (hPa) per 5° of altitude? What is the contour gradient in metres per 5° of altitude?

2. If an upper level jet at 250 mb is 200 knots and contours are spaced 60 m apart what is the distance between two contours in km?

14

FRICTION IN THE BOUNDARY LAYER OF THE ATMOSPHERE

14.1 INTRODUCTION

In all the equations we have derived so far we have assumed that the atmosphere was frictionless. This is a reasonably valid assumption to make through much of the atmosphere. However, in the planetary boundary layer, which may extend to only 30 m above the ground in very stable conditions, and to as much as 3 km above the ground in turbulent or convective conditions, the motion of the air is obstructed by surface friction. This may be due to the roughness of the land surface due to vegetation, trees, small hills or mountains. The rigorous way of treating friction in the planetary boundary layer is by means of turbulence theory. This is a highly complex subject and we shall only touch on its simplest aspects in this work. We shall first, however, consider friction in a simpler way.

14.2 THE GULDBERG-MOHN APPROXIMATION

Guldberg–Mohn (1876) assumed that surface friction could be parameterized by letting the frictional deceleration of the air be proportional to the wind velocity. This is true to a first approximation, although it is not based on a rigid physical theory. However, the treatment of friction in this way does allow the equations of motion to be treated in a relatively simple manner.

Thus we may write equations (7.4) as

$$\frac{du}{dt} - fv = -\frac{1}{\rho}\frac{\partial p}{\partial x} - ku$$

$$\frac{dv}{dt} + fu = -\frac{1}{\rho}\frac{\partial p}{\partial y} - kv$$

$$(14.1)$$

where k is a constant of proportionality. This constant has been experimentally found to vary from about $1.0 \times 10^{-5}\,\text{s}^{-1}$ over a fairly calm sea to about 10 times as much over a land surface covered by grass or similar vegetation. The frictional force acts to reinforce the pressure gradient. It slows down the wind velocity so that it is no longer in balance with the pressure field. The result is that the wind flows across the isobars from high to low pressure.

14.3 BALANCED FRICTIONAL FLOW

Let us suppose that the flow is balanced for a system of straight isobars orientated from east to west along the x axis. In this case du/dt and dv/dt in equations (14.1) are equal to zero, since there is no acceleration. We have instead

$$v = \frac{ku}{f}$$

$$u = \frac{-1/\rho}{f}\left(\frac{\partial p}{\partial y}\right) - \frac{kv}{f}$$

and the two components u, v may be written

$$u = \frac{f^2 u_g}{(f^2 + k^2)}$$

$$v = \frac{kfu_g}{(f^2 + k^2)}$$

(14.2)

$$V = \sqrt{u^2 + v^2} = \frac{fu_g}{(f^2 + k^2)}$$

(14.3)

where V is the total wind velocity. If θ is the angle between the total wind direction and the isobars, then $\tan\theta = v/u = k/f$.

Figure 14.1 illustrates the balance of forces for the case discussed.

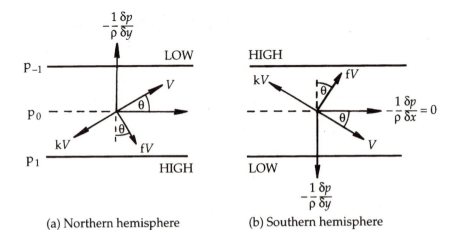

(a) Northern hemisphere   (b) Southern hemisphere

FIGURE 14.1 Inclination of the wind across the isobars as a result of friction.

14.4 THE NEWTONIAN CONCEPT OF FRICTION

Although the Guldberg–Mohn parameterization of friction is extremely convenient in some models, particularly in Lagrangian representations of motion, it
does not embody the rigid mathematical physics of an exact explanation of the
effect of friction. To do this we must venture into the field of turbulence, and
consider that friction is due to the effects of eddy viscosity.

Let us consider a layer of air. Its lower boundary is the surface and its upper
boundary is some higher level where the effects of surface friction are less evident
(*see* Fig. 14.2(a)).

At the upper level the wind has a velocity $u(z)$ and at the lower boundary
the velocity is zero, while the distance separating the two levels is l. Then the
tangential shearing stress exerted by one infinitesimal layer on its adjacent layer
may be defined as

$$\tau_x = \mu \frac{\partial u}{\partial z} \tag{14.4}$$

where μ is a viscosity coefficient which is assumed to be constant. The stress is
tangential to the x axis. The schematic concept illustrated in Fig. 14.2 is said to

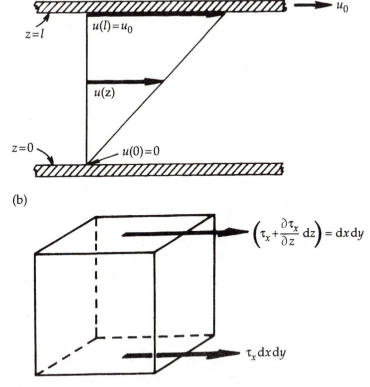

FIGURE 14.2 The shearing stress force.

have been predicted by Newton to describe molecular processes. The usual example shown in textbooks is to show a shear of the fluid between an upper and a lower bounding plate, but the process may be transformed so that it represents eddy viscosity in the atmosphere. μ then becomes a coefficient of eddy viscosity. The subscript in (14.4) indicates that the shearing stress is in the x direction and due to the vertical wind shear. Physically, it represents an eddy transfer of momentum through the interface of two adjacent layers. Now if there is a variation of stress between two infinitesimal layers there will be a divergence of shearing stress and a force per unit mass or acceleration will be exerted by a layer on its neighbouring layer. These forces may be regarded as the transfer of momentum from faster moving air at higher levels to slower moving air at lower levels, and vice versa. This interchange of momentum between levels may then be interpreted as a frictional retardation of the flow which would otherwise occur if there were no reduction of velocity by the lower boundary.

We may now derive a term which expresses the acceleration (or retardation) of the flow due to the variation with height of the shearing stress.

Consider the rectangular box in Fig. 14.2(b). The drag exerted in the x direction on the lower face $dx\,dy$ is $\tau_x\,dx\,dy$; the drag on the upper face is

$$\left(\tau_x + \frac{\partial \tau_x}{\partial z}\,dz\right)dx\,dy$$

Then the difference between these two forces is the shearing stress force acting on the volume element

$$\left(\tau_x + \frac{\partial \tau_x}{\partial z}\,dz\right)dx\,dy - \tau_x dx\,dy = \frac{\partial \tau_x}{\partial z}\,dV = \frac{1}{\rho}\frac{\partial \tau_x}{\partial z} \quad \text{for unit mass} \quad (14.5)$$

Similarly one can express the shearing stress force along the y axis. It is

$$\frac{1}{\rho}\frac{\partial \tau_y}{\partial z}$$

If equations (14.4) and (14.5) are combined, it follows that

$$\frac{1}{\rho}\frac{\partial \tau_x}{\partial z} = \frac{\mu}{\rho}\frac{\partial^2 u}{\partial z^2} \qquad \frac{1}{\rho}\frac{\partial \tau_y}{\partial z} = \frac{\mu}{\rho}\frac{\partial^2 v}{\partial z^2} \qquad (14.6)$$

The expressions in (14.6) may now be included in the equations of motion:

$$\frac{du}{dt} - fv = \frac{-1}{\rho}\frac{\partial p}{\partial x} + \frac{1}{\rho}\frac{\partial \tau_{zx}}{\partial z}$$

$$\frac{dv}{dt} + fu = \frac{-1}{\rho}\frac{\partial p}{\partial y} + \frac{1}{\rho}\frac{\partial \tau_{zy}}{\partial z}$$

$$(14.7)$$

We may visualize the physical meaning of the final terms from a meteorological point of view. When the wind is blowing over a solid land surface (or liquid ocean surface) the stress will be greatest at that boundary and will decrease until it becomes a minimum at some upper level, normally the level at which the wind is assumed to be geostrophic. The stress therefore decreases with height. That means the shear of the wind decreases with height. Thus the terms contained in (14.6) are negative and the air flow is retarded. We are especially interested in a balanced state in which the flow is steady and does not possess any acceleration. Under such conditions equations (14.7) reduce to

$$K \frac{\partial^2 u}{\partial z^2} + fv' = 0$$

$$K \frac{\partial^2 v}{\partial z^2} - fu' = 0$$

(14.8)

In (14.8) u', v' are the geostrophic departures and $K = \mu/\rho$, the eddy coefficient of viscosity, which is assumed to be constant with height. Equations (14.8) are usually applied to what is called the Ekman or spiral layer within the planetary boundary layer. The lower boundary of the Ekman layer is normally assumed to be about 10 metres above the actual surface. The layer between the surface and 10 metres is governed by a different process. This layer is called the surface layer.

14.5 THE SURFACE LAYER

The viscosity force in equations (14.7) may be expressed in the form $\partial(\tau/\rho)/\partial z$. In dealing with the surface layer we define a new and very important parameter called the friction velocity. It is denoted u^*, where

$$u^* = \sqrt{\frac{\tau}{\rho}}$$

(14.9)

If a parcel of air is displaced vertically it is assumed that $u' = kz(\partial \bar{u}/\partial z)$ where u' is a perturbed velocity from $\bar{u}$, the mean flow, and $kz = l$, some typical mixing length of eddy. Then, if we consider the motion along the x axis we may express the eddy shearing stress, as before,

$$\tau = \mu \frac{\partial u}{\partial z}$$

and

$$\mu = \rho k^2 z^2 \frac{\partial u}{\partial z}$$

(14.10)

Equation (14.10) is derived from what is called mixing length theory. This concept is discussed at length in more advanced textbooks. We note that μ is no longer constant as it was assumed when we initially defined stress for the case of uniform motion without eddies. It is now the eddy exchange coefficient

$$\frac{\tau}{\rho} = k^2 z^2 \left(\frac{\partial \bar{u}}{\partial z} \right)^2$$

$$\tau_{xz} = \rho k^2 z^2 \left(\frac{\partial u}{\partial z} \right)^2$$

(14.11)

From (14.11), the definition of the friction velocity is

$$u^* = kz \frac{\partial u}{\partial z}$$

Integrating,

$$u = \frac{u^*}{k} \frac{\log z}{z_0} \tag{14.12}$$

where z_0 is a roughness coefficient dependent on the roughness of the surface to fit the condition that $u = 0$ at $z = z_0$. k is called Von Karman's constant and is equal to about 0.4. Thus, equation (14.12) describes the velocity profile in the surface layer between the surface and about 10 metres height. Above the surface layer we enter the Ekman or spiral layer which is governed by equations (14.8).

14.6 THE SPIRAL OR EKMAN LAYER

We will now derive the equations which express the wind components as a function of height above some reference level as a function of the height. To do this we must solve equations (14.8). To simplify the problem we will orientate the isobars along the x axis as we have done previously in Chapter 10. Equations (14.8) then becomes

$$K \frac{\partial^2 u}{\partial z^2} + fv = 0$$
$$K \frac{\partial^2 v}{\partial z^2} - fu' = 0 \tag{14.13}$$

We may solve (14.13) for the appropriate boundary conditions in two ways. The first is to obtain the constants of integration by orthodox algebraic manipulation. The second is to introduce the concept of complex numbers, which involve the square root of -1, which we call i. We will here use the more conventional method, and leave the second, more sophisticated method to be worked out under some of the worked examples of this chapter. In both cases we have to rely on standard-type solutions. We used this method in solving a set of second-order differential equations in Chapter 10. In that case equations (10.3) were total differential equations whereas we have now have a set of differential equations written in their partial form. However, since we are only interested in the variation in the vertical (the arbitrary constants are only functions of the reference horizontal plane), we may solve them as if they were ordinary differential equations.

A standard-type solution of equations (14.13) is given by $u' = A$. We will write a standard solution of (14.13) in the form

$$u = Au_g \, e^{-az} \sin(az - b)$$
$$v = Bu_g \, e^{-az} \cos(az - b)$$

where $u' = u - u_g$. Now friction slows the motion down so that u' will be negative, but u will be positive if the isobars are orientated along the x axis. v will blow across the isobars from high to low pressure and will be positive, since pressure decreases along the y axis. We have to evaluate the constants A, B, a, b. To do this we differentiate the standard solutions partially twice with respect to z.

Then

$$\frac{\partial u}{\partial z} = -Aau_g e^{-az}[\sin(az - b) - \cos(az - b)]$$

$$\frac{\partial^2 u}{\partial z^2} = -2Aa^2 u_g e^{-az} \cos(az - b)$$

and

$$\frac{\partial v}{\partial z} = -Bau_g e^{-az}[\cos(az - b) + \sin(az - b)]$$

$$\frac{\partial^2 v}{\partial z^2} = -2Ba^2 u_g e^{-az} \sin(az - b)$$

Substituting the second partial derivatives above in (14.13) for the boundary conditions $u_0 = 0$, $v_0 = 0$, $z_0 = 0$ we obtain

$$2KAa^2 = fB$$

and

$$2KBa^2 = fA$$

We see that $A^2 = B^2$ or $A = \pm B$.

We now return to the standard solution and substitute the boundary conditions. We obtain $A \sin b = 1$ and $Bu_g \cos b = 0$.

The equations (14.13) are therefore satisfied if $A = 1$, $B = 1$, and $b = \pi/2$.

Since $\sin(\phi - \pi/2) = -\cos\phi$ and $\cos(\phi - \pi/2) = \sin\phi$ our Ekman spiral equations are

$$u = u_g(1 - e^{-az} \cos az)$$
$$v = u_g e^{-az} \sin az \qquad (14.14)$$

We may interpret (14.14) in the sense that as the elevation above the surface increases indefinitely $u \to u_g$ and $v \to 0$. The gradient wind level, that is the level at which the wind becomes quasi-geostrophic, is about 500 metres or so above the surface. Figure 14.3 shows a plot of the spiral. It is seen that the wind along the isobars actually becomes supergeostrophic at higher levels. It is only when z becomes very large that u becomes exactly geostrophic.

Equations (14.14) have been derived under the premise that the spiral layer extends down to the surface. This is not truly so. We have previously discussed the structure of the surface layer which may be said to describe the first 10 metres of the atmosphere above the earth's surface. The latter structure offers a much closer approximation to the truth for the surface layer. We may therefore model a more accurate representation of the spiral layer if we place its lower boundary at the top of the surface layer, about 10 metres above the surface. Standard anemometers are positioned at about this height. If we make this condition we need a new lower boundary condition. Such a condition may be satisfied if we

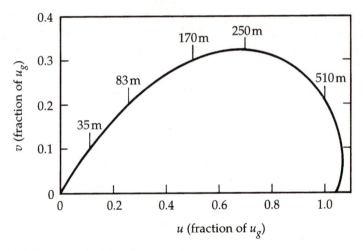

FIGURE 14.3 Ekman spiral for the case where the surface level is assumed to be at anemometer level. The angle of inclination of the wind across the isobars at surface level is 15°.

make the assumption that the wind at the new $z = 0$ level (actually 10 metres above the surface) is proportional to the shear. Then,

$$u_0 = c\left(\frac{\partial u}{\partial z}\right)_0 \qquad v_0 = c\left(\frac{\partial v}{\partial z}\right)_0$$

where c is a constant of proportionality. At the boundary

$$\left(\frac{\partial u}{\partial z}\right)_0 = \frac{u_0}{c} = Aau_g(\sin b + \cos b)$$

$$\left(\frac{\partial v}{\partial z}\right)_0 = \frac{v_0}{c} = Bau_g(\sin b - \cos b) \qquad (14.15)$$

$$\tan \alpha_0 = \frac{v_0}{u_0} = \frac{B\sin b - \cos b}{A \sin b + \cos b}$$

Evaluating the constants by substitution of the lower boundary values into (14.14) and (14.15) above

$$u_0 = u_g + Au_g \sin b = cAau_g(\sin b + \cos b)$$

$$v_0 = Bu_g \cos b = cBau_g(\sin b - \cos b)$$

Then from the second relation, that is for v_0

$$\tan b = \frac{1 + ac}{ac} \qquad (14.16)$$

and from the first relation, that is for u_0,

$$A = \frac{1}{\sqrt{1 + 2ac + 2a^2c^2}}$$

Worked Example

Show that

$$A^2 = \frac{1}{1 + 2ac + 2a^2 c^2}$$

Solution:

We have already found that

$$\tan b = \frac{(1 + ac)}{ac}$$

From the boundary conditions

$$1 - A \sin b = cAa(\sin b + \cos b)$$

$$A = \frac{1}{\sin b + ac \sin b + ac \cos b}$$

Then from right-angled triangle relationships

$$\sin b = \frac{1 + ac}{\sqrt{1 + 2ac + 2a^2 c^2}}$$

$$\cos b = \frac{ac}{\sqrt{1 + 2ac + 2a^2 c^2}}$$

Therefore

$$A^2 = \frac{1}{(1 + 2ac + 2a^2 c^2)}$$

We already know from the derivation of the classical Ekman spiral that $A = B$.
Then

$$\tan \alpha = \frac{\sin b - \cos b}{\sin b + \cos b}$$

$$= \frac{\tan b - 1}{\tan b + 1} = \frac{1}{1 + 2ac}$$

where α is the angle between the wind and the isobars at the lower boundary
level, that is at anemometer height level.

We know that $a^2 = f/2k$ since $A = B$. Thus, if $f = 0$, $a = 0$, and $\alpha = 45°$.
Thus at the equator the wind would blow across the isobars at an angle of $45°$.
Also it can be shown that $A^2 = 2 \sin^2 \alpha$.

We know from (14.15) that

$$\tan \alpha = \frac{\tan b - 1}{\tan b + 1} = \tan\left(b - \frac{\pi}{4}\right)$$

where $\alpha = b - \pi/4$ or $b = \alpha + \pi/4$. Then, having evaluated A and b, we can write

$$u = u_g\left[1 - \sqrt{2} \sin \alpha\, e^{-az} \cos\left(az - \alpha + \frac{\pi}{4}\right)\right]$$

$$v = u_g\, e^{-az} \sqrt{2} \sin \alpha \sin\left(az - \alpha + \frac{\pi}{4}\right)$$

(14.17)

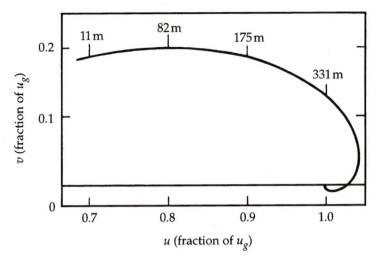

FIGURE 14.4 Modified Ekman spiral.

Figure 14.4 shows a plot of equations (14.17) which is sometimes referred to as the modified Ekman spiral in contrast to the simpler classical spiral depicted in Fig. 14.3. It can readily be seen that the two versions become identical when $\alpha = \pi/4$. A useful application of the Ekman spiral is the calculation of the vertical velocity within a pattern of surface isobars which possess cyclonic vorticity. Such cyclonic vorticity may be cyclonic shear in an east-to-west isobar (high-pressure to the south in the northern hemisphere and low-pressure to the south in the southern hemisphere).

Worked Example
The synoptic chart on a given day shows a pattern of east–west-orientated isobars with the geostrophic wind decreasing with latitude in the direction of low-pressure. Show that the vertical velocity at the top of the frictional layer is proportional to the geostrophic vorticity. Use the classical Ekman spiral.

Solution:

$$\zeta_g = -\frac{\partial u_g}{\partial y}$$

$$\frac{\partial u}{\partial x} + \frac{\partial v}{\partial y} = -\frac{\partial w}{\partial z}$$

$$\frac{\partial u}{\partial x} = 0 \tag{14.18}$$

$$\frac{\partial w}{\partial z} = -\frac{\partial v_g}{\partial y}$$

$$\Delta w = \int_0^\infty \frac{\partial w}{\partial z}\, dz = -\int_0^\infty \frac{\partial v_g}{\partial y}\, dz$$

From the Ekman solution, assuming $w = 0$ at the surface,

$$w = \int_0^\infty \zeta_g (e^{-az} \sin az) \, dz$$

(14.19)

$$w = \frac{\zeta_g}{2a}$$

Worked Example
If $\zeta_g = 10^{-5}\,\mathrm{s}^{-1}, f = 10^{-4}\,\mathrm{s}^{-1}$ and $k = 10\,\mathrm{m}^2\,\mathrm{s}^{-1}$, what is the numerical value of the vertical velocity?

Solution:

$$a = \sqrt{\frac{f}{2k}}$$

(14.20)

$$w = \frac{\zeta_g \sqrt{k}}{\sqrt{2f}} = 0.22\,\mathrm{cm\,s}^{-1}$$

Worked Example
Show that the ratio of the speeds of the surface wind to the geostrophic wind is $v_0/v_g = \cos\alpha - \sin\alpha$ using equations (14.17) for the lower boundary.

Solution:

$$u_0 = u_g \left[1 - \sqrt{2} \sin\alpha \cos\left(\frac{\pi}{4} - \alpha\right) \right]$$

(14.21)

$$v_0 = u_g \sqrt{2} \sin\alpha \cos\left(\frac{\pi}{4} - \alpha\right)$$

(14.22)

Then

$$\frac{u_0^2 + v_0^2}{u_g^2} = 1 - 2\sqrt{2} \sin\alpha \cos\left(\frac{\pi}{4} - \alpha\right) + 2\sin^2\alpha \cos^2\left(\frac{\pi}{4} - \alpha\right)$$

$$+ 2\sin^2\alpha \cos^2\left(\frac{\pi}{4} - \alpha\right)$$

(14.23)

$$\frac{v_0^2}{v_g^2} = 1 - 2\sqrt{2} \sin\alpha \cos\left(\frac{\pi}{4} - \alpha\right) + 2\sin^2\alpha$$

$$= 1 - 2\sqrt{2} \sin\alpha \left(\frac{\sqrt{2}}{2} \cos\alpha + \frac{\sqrt{2}}{2} \sin\alpha \right) + 2\sin^2\alpha$$

$$= 1 - 2\sin\alpha \cos\alpha - 2\sin^2\alpha + 2\sin^2\alpha$$

(14.24)

or

$$\frac{v_0^2}{v_g^2} = 1 - 2\sin\alpha \cos\alpha = \sin^2\alpha + \cos^2\alpha - 2\sin\alpha \cos\alpha$$

and thus

$$\frac{v_0}{v_g} = \cos\alpha - \sin\alpha$$

(14.25)

The Ekman spiral may also be found by resorting to complex numbers. To do this we multiply the v component of the wind by i, the square root of -1. Then from (14.13)

$$\frac{k\partial^2(u+iv)}{\partial z^2} + fv - if(u - ug) = 0 \qquad (14.26)$$

$$\frac{k\partial^2(u+iv)}{\partial z^2} - if(u+iv) + ifu_g = 0 \qquad (14.27)$$

The general solution is

$$u + iv = A\exp\left[\left(\frac{if}{k}\right)^{1/2}z\right] + B\exp\left[-\left(\frac{af}{k}\right)^{1/2}\right] + u_g$$

$$\frac{\partial^2(u+iv)}{\partial z^2} = A\left[\frac{if}{k}\exp\left(-\frac{if}{k}\right)^{1/2}\right] + B\left[\frac{if}{k}\exp\left(-\frac{if}{k}\right)^{1/2}\right] \qquad (14.28)$$

Now

$$\sqrt{i} = \frac{1+i}{\sqrt{2}} \qquad a = \sqrt{\frac{f}{2k}} \qquad (14.29)$$

It is obvious that $A = 0$ since the velocity cannot increase indefinitely with height. Inserting the boundary conditions for the classical case, $u_0 = 0$, $v_0 = 0$ at $z = 0$,

$$u + iv = -u_g\left[e^{-(a+ai)z} + u_g\right]$$

$$= u_g\left(1 - e^{-az-iaz}\right) \qquad (14.30)$$

but

$$e^{-iaz} = \cos az - i\sin az$$

so

$$u + iv = u_g[1 - e^{-az}(\cos az - i\sin az)] \qquad (14.31)$$

Matching the real and imaginary parts

$$u = u_g(1 - e^{-az}\cos az)$$

$$v = u_g e^{-az}\sin az \qquad (14.32)$$

as before.

14.7 PROBLEMS

1. A wide river flows 1000 km along the equator to the sea. At 1000 km from the sea its elevation above m.s.l. is 1000 m. The slope is constant and the downstream velocity is $1\,\mathrm{m\,s}^{-1}$. Assuming the frictional retardation of the flow is proportional to the velocity, what is the numerical value of the constant of frictional resistance? Suppose the downward slope is from south to north between 40° and 60° of latitude and the northward velocity of the river is $1\,\mathrm{m\,s}^{-1}$. Does the value of k alter?

2. Let us assume the Guldberg–Mohn parameterization of friction. The pressure gradient is constant. What is the relation between k and f for the maximum cross-isobaric wind to occur at any given latitude?

3. Assuming the classical Ekman spiral, at what altitude is the cross-isobaric velocity a maximum? Assume constant geostrophic wind.

4. Suppose the geostrophic wind increases with height so that $u_g = u_{g_0} + Cz$. Compute the Ekman spiral for the first 500 metres if (a) the geostrophic velocity at 500 metres is double that at the surface, (b) it is half that at the surface.

15

SOME MORE ADVANCED EQUATIONS

In Chapter 11 we derived the vorticity equation by cross-differentiating the equations of motion in horizontal, frictionless flow. By cross-differentiation we mean we form $\partial/\partial y$ of the first equation and $\partial/\partial x$ of the second equation. In doing this and subtracting we were able to eliminate the pressure gradient term and derive a relation characteristic of the motion itself. We will now differentiate in x, y, ignoring horizontal variations in the vertical motion and density.

We then obtain

$$\frac{\partial^2 u}{\partial x^2} + u\frac{\partial^2 u}{\partial x^2} + \left(\frac{\partial u}{\partial x}\right)^2 + v\frac{\partial^2 u}{\partial x\,\partial y} + \frac{\partial u}{\partial y}\frac{\partial v}{\partial x} - f\frac{\partial v}{\partial x} = -\frac{1}{e}\frac{\partial^2 p}{\partial t^2} \qquad (15.1)$$

$$\frac{\partial^2 v}{\partial y\,\partial t} + u\frac{\partial^2 v}{\partial x\,\partial y} + \frac{\partial u}{\partial y}\frac{\partial v}{\partial x} + \left(\frac{\partial v}{\partial y}\right)^2 + v\frac{\partial^2 v}{\partial y^2} + f\frac{\partial u}{\partial y} + u\frac{\partial f}{\partial y} = -\frac{1}{e}\frac{\partial^2 p}{\partial y^2} \qquad (15.2)$$

We will add the two equations above. On the right hand side we will replace $-(1/\rho)\nabla^2 p$ by $\nabla^2\phi$ where $\phi = -g\nabla z$ the geopotential gradient. Collecting terms we have

$$\frac{\partial}{\partial t}\left(\frac{\partial u}{\partial x} + \frac{\partial v}{\partial y}\right) - f\left(\frac{\partial v}{\partial x} - \frac{\partial u}{\partial y}\right) + u\frac{\partial f}{\partial y} + \left(\frac{\partial u}{\partial x} + \frac{\partial v}{\partial y}\right)^2$$

$$+ u\frac{\partial}{\partial x}\left(\frac{\partial u}{\partial x} + \frac{\partial v}{\partial y}\right) + v\frac{\partial}{\partial y}\left(\frac{\partial u}{\partial x} + \frac{\partial v}{\partial y}\right)$$

$$- 2\frac{\partial u}{\partial x}\frac{\partial v}{\partial y} + 2\frac{\partial u}{\partial y}\frac{\partial v}{\partial x} = -\nabla^2\phi \qquad (15.3)$$

Letting

$$\frac{\partial u}{\partial x} + \frac{\partial v}{\partial y} = D \qquad \frac{\partial v}{\partial x} - \frac{\partial u}{\partial y} = \zeta \quad \text{and} \quad \beta = \frac{\partial f}{\partial y}$$

and simplifying we have

$$\frac{dD}{dt} + D^2 - f\zeta + \beta u - 2J(u, v) = -\nabla^2 \phi \qquad (15.4)$$

Equation (15.3) is known as the divergence equation. We may now make some approximations.

1. If $\partial D/\partial t = 0$ the divergence is in a steady state but the divergence will change in terms of the spatial pattern of the contours.
2. If $\partial D/\partial t = 0$, the divergence within the flow will remain constant within the domain for which the equation is applicable.
3. If $\partial D/\partial t = 0$ and $D = 0$ the flow will be non-divergent.

For case 1 the equation eliminates time-dependent gravity waves. For case 2 the equation eliminates both time-dependent and standing gravity waves.

15.2 THE BALANCE EQUATION

For case 3 the equation describes the relation between the flow of the wind and the isobars or contours for the non-divergent wind. It is, in fact, an extension of the gradient wind equation to cover the continuous curving of the isobars across the weather map.

We have

$$\beta u - f\zeta - 2J(u, v) + \nabla^2 \phi = 0 \qquad (15.5)$$

Since the wind is non-divergent we may transform it into a stream function using the relations

$$u = -\frac{\partial \psi}{\partial y} \qquad v = \frac{\partial \psi}{\partial x}$$

and obtain the balance equation

$$\nabla^2 \psi - f\nabla^2 \psi - \beta \frac{\partial \psi}{\partial y} + 2\left[\left(\frac{\partial^2 \psi}{\partial x \, \partial y}\right)^2 - \frac{\partial^2 \psi}{\partial x^2}\frac{\partial^2 \psi}{\partial y^2}\right] = 0 \qquad (15.6)$$

Worked Example
Show that $\beta u - f\zeta = -\nabla \cdot (f\nabla \psi)$.

Solution:

$$u = -\frac{\partial \psi}{\partial y}$$

$$\zeta = \nabla^2 \psi$$

$$\beta u - f\zeta = -\beta \frac{\partial \psi}{\partial y} - f\nabla^2 \psi$$

$$-\nabla \cdot (f\nabla \psi) = -f\nabla \nabla \psi - \left(\mathbf{i}\frac{\partial \psi}{\partial x} + \mathbf{j}\frac{\partial \psi}{\partial y}\right) \cdot \nabla f$$

$$-\nabla \cdot (f\,\nabla\psi) = -f\,\nabla^2\psi - \left(\mathbf{i}\frac{\partial\psi}{\partial x} + \mathbf{j}\frac{\partial\psi}{\partial y}\right) \cdot \left(\mathbf{i}\frac{\partial f}{\partial x} + \mathbf{j}\frac{\partial f}{\partial y}\right)$$

$$-\nabla \cdot (f\,\nabla\psi) = -f\,\nabla^2\psi - \beta\frac{\partial\psi}{\partial y}$$

Equation (15.3) may be used to determine the initial wind for numerical prediction models. It is too cumbersome an equation to solve for ordinary operational weather forecasting purposes. Although more exact than the geostrophic wind, the latter serves a close enough approximation for normal, routine synoptic analyses.

15.3 THE OMEGA EQUATION

We will derive one final equation which expresses a comprehensive behaviour of the motion of the atmosphere. It relates the vertical motion to the existing geopotential height field. In deriving this important equation we will review all the basic equations upon which this ultimate relation depends.

The equations of horizontal, frictionless motion including the vertical motion in pressure coordinates may be written as

$$\frac{\partial u}{\partial t} + u\frac{\partial u}{\partial x} + v\frac{\partial u}{\partial y} + \omega\frac{\partial u}{\partial p} - fv = -\frac{\partial\Phi}{\partial x}$$

$$\frac{\partial v}{\partial t} + u\frac{\partial v}{\partial x} + v\frac{\partial v}{\partial y} + \omega\frac{\partial v}{\partial p} + fu = -\frac{\partial\Phi}{\partial x} \qquad (15.7)$$

where ω is defined as $\mathrm{d}p/\mathrm{d}t$.

The hydrostatic equation is $\partial p/\partial z = -\rho g$, but must be written in inverse form as

$$\frac{\partial z}{\partial p} = -\frac{\alpha}{g}$$

$$g\frac{\partial z}{\partial p} = -\alpha \qquad (15.8)$$

$$\frac{\partial\Phi}{\partial p} = -\alpha = -\frac{RT}{p}$$

where Φ is the geopotential height of the pressure surface.

The compressible continuity equation in pressure coordinates is

$$\frac{\partial u}{\partial x} + \frac{\partial v}{\partial y} + \frac{\partial\omega}{\partial p} = 0$$

The thermodynamic equation is

$$C_p\,\mathrm{d}T - \alpha\,\mathrm{d}p = q$$

from which, assuming the motion is adiabatic,

$$C_p\left(\frac{\partial T}{\partial t}+u\frac{\partial T}{\partial x}+v\frac{\partial T}{\partial y}+\omega\frac{\partial T}{\partial p}\right)-\alpha\omega=\dot{q}=0$$

$$\frac{\partial T}{\partial t}+u\frac{\partial T}{\partial x}+v\frac{\partial T}{\partial y}+\frac{\omega\partial T}{-\rho g\partial z}-\frac{\alpha\omega}{C_p}=0$$

$$\frac{\partial T}{\partial t}+u\frac{\partial T}{\partial x}+v\frac{\partial T}{\partial y}+\frac{\omega\gamma}{\rho g}-\frac{\alpha\omega}{g}\Gamma_d=0$$

$$\frac{\partial T}{\partial t}+u\frac{\partial T}{\partial x}+v\frac{\partial T}{\partial y}+\omega\gamma\left(-\frac{\partial z}{\partial p}\right)+\frac{\partial z}{\partial p}\omega\Gamma_d$$

$$\frac{\partial T}{\partial t}+u\frac{\partial T}{\partial x}+v\frac{\partial T}{\partial y}+\frac{\partial z}{\partial p}(\Gamma_d-\gamma)\omega=0$$

Let

$$\frac{\partial z}{\partial p}(\Gamma_d-\gamma)=-S_p$$

where S_p is a stability parameter. Then

$$\frac{\partial T}{\partial t}+u\frac{\partial T}{\partial x}+v\frac{\partial T}{\partial y}-S_p\omega=0 \tag{15.9}$$

We also know that

$$\frac{d\theta}{\theta}=\frac{dT}{T}-\kappa\frac{dp}{p}$$

$$\frac{1}{\theta}\frac{d\theta}{dz}=\frac{1}{T}\frac{dT}{dz}-\frac{k}{p}\frac{dp}{dz}=-\frac{\gamma}{T}+\frac{\Gamma_d}{T}$$

$$\frac{T}{\theta}\frac{\partial\theta}{\partial z}=\Gamma_d-\gamma \tag{15.10}$$

$$\frac{T}{\theta}\frac{\partial\theta}{\partial z}\frac{dz}{dp}=T\frac{\partial}{\partial p}\log\theta=(\Gamma_d-\gamma)\frac{\partial z}{\partial p}=-S_p$$

where $\kappa=R/c_p$.

We define a static stability parameter

$$\sigma=\frac{-\alpha\partial\theta}{\theta\partial p}=\frac{-RT}{\theta p}\frac{\partial\theta}{\partial p}=\frac{R}{p}S_p$$

Now if $\gamma<\Gamma_d$ the lapse rate is stable and σ is positive. Alternatively if $\gamma>\Gamma_d$ the lapse rate is unstable and σ is negative. Except in regions of strong convection, σ is normally positive. As shown in problem 2, σ is defined by the vertical structure of the geopotential height field.

We will now make a number of approximations in deriving the omega equation by using the vorticity equation.

1. Assume geostrophic flow for the advection term.
2. Neglect the tilting and twisting terms in the geostrophic vorticity equation.

3. Consider that $f = f_0$, a constant value for the domain covered by each calculation. It depends on the latitude of the domain.
4. Consider $\beta = \partial f / \partial y$ a constant at the latitude of the domain.
5. Neglect ζ compared with f in the divergence term of the vorticity equation. Note that the horizontal wind in the divergence term is not replaced by the geostrophic velocity. If it were, the divergence would be zero and no vertical motion would be generated. Instead we use the continuity equation $\text{div } \mathbf{V} = -\partial w / \partial p$.

Having made the assumptions we write the vorticity equation as

$$\frac{\partial \zeta_g}{\partial t} + \frac{u_g \partial \zeta_g}{\partial x} + \frac{v_g \partial \zeta_g}{\partial y} + \beta v_g = f_0 \frac{\partial w}{\partial p} \tag{15.11}$$

noting that $\zeta_g = \nabla^2 \Phi / f_0$.

We will rewrite (15.9) using the hydrostatic relation (15.8) and noting that $S_p = p\sigma / R$:

$$\frac{\partial}{\partial t}\left(-\frac{\partial \Phi}{\partial p}\right) = u_g \frac{\partial}{\partial x}\left(-\frac{\partial \Phi}{\partial p}\right) + v_g \frac{\partial}{\partial y}\left(-\frac{\partial \Phi}{\partial p}\right) + \sigma w = 0 \tag{15.12}$$

Replacing $\partial \Phi / \partial t$ by χ we have

$$\frac{\partial \chi}{\partial p} + u_g \frac{\partial}{\partial x}\left(\frac{\partial \Phi}{\partial p}\right) + v_g \frac{\partial}{\partial y}\left(\frac{\partial \Phi}{\partial p}\right) + \sigma w = 0 \tag{15.13}$$

and in vector form

$$\frac{\partial \chi}{\partial p} + \mathbf{V}_g \cdot \nabla \frac{\partial \Phi}{\partial p} + \sigma w = 0 \tag{15.14}$$

We will now rewrite equation (15.11) as

$$\frac{1}{f_0}\frac{\partial \nabla^2 \Phi}{\partial t} + \frac{1}{f_0}\left(u_g \frac{\partial \nabla^2 \Phi}{\partial x} + v_g \frac{\partial \nabla^2 \Phi}{\partial y}\right) + \beta v_g = f_0 \frac{\partial w}{\partial p}$$

$$\nabla^2 \frac{\partial \Phi}{\partial t} + f_0\left(\frac{1}{f_0}u_g \nabla^2 \frac{\partial \Phi}{\partial x} + \frac{1}{f_0}v_g \nabla^2 \frac{\partial \Phi}{\partial y}\right) + f_0 \beta v_g = f_0^2 \frac{\partial w}{\partial p} \tag{15.15}$$

$$\nabla^2 \chi = -u_g \nabla^2 \frac{\partial \Phi}{\partial x} - v_g \nabla^2 \frac{\partial \Phi}{\partial y} + f_0 \beta v_g + f_0^2 \frac{\partial w}{\partial p}$$

or in vector form, involving the simple dot product,

$$\nabla^2 \chi = -f_0 \mathbf{V}_g \cdot \nabla \left(\frac{\nabla^2 \Phi}{f_0} + f\right) + f_0^2 \frac{\partial w}{\partial p} \tag{15.16}$$

We wish to eliminate χ between (15.14) and (15.16). To do this we take the Laplacian of (15.14):

$$\nabla^2 \frac{\partial \chi}{\partial p} + \nabla^2 \left(\mathbf{V}_g \cdot \nabla \frac{\partial \Phi}{\partial p}\right) + \sigma \nabla^2 w = 0 \tag{15.17}$$

We then take $\partial/\partial p$ of (15.16):

$$\nabla^2 \frac{\partial \chi}{\partial p} + f_0 \frac{\partial}{\partial p}\left[\mathbf{V}_g \cdot \nabla\left(\frac{\nabla^2 \Phi}{f_0} + f\right)\right] - f_0^2 \frac{\partial^2 w}{\partial p^2} = 0 \qquad (15.18)$$

Subtracting (15.18) from (15.17) we obtain, after dividing by σ,

$$\nabla^2 w + \frac{f_0^2}{\sigma}\frac{\partial^2 w}{\partial p^2} + \frac{1}{\sigma}\nabla^2\left[\mathbf{V}_g \cdot \nabla \frac{\partial \Phi}{\partial p}\right] - \frac{f_0}{\sigma}\frac{\partial}{\partial p}\left[\mathbf{V}_g \cdot \nabla\left(\frac{\nabla^2 \Phi}{f_0} + f\right)\right] = 0 \quad (15.19)$$

Equation (15.19) is called the geostrophic omega equation. It relates the geopotential heights, that is the contours on a synoptic chart, with the vertical velocity field. Usually, it must be evaluated by computer. When the water vapour structure of the atmosphere is compared with the vertical motion, maps of rainfall intensities and amounts may be prepared.

But perhaps even more important than the vertical velocity as an output of the contour pattern is the remarkable way in which the atmosphere's behaviour is regulated and controlled in accordance with the basic assumptions of geostrophic flow, hydrostatic equilibrium and mass conservation.

15.4 PROBLEMS

1. Show that the balance equation (15.6) can be written as

$$\nabla^2\left[\Phi + \tfrac{1}{2}(\nabla\Phi)^2\right] - \nabla[(f + \nabla^2\Phi)\nabla\Phi] = 0$$

2. Show that the static stability parameter σ may be written in terms of Φ in the form

$$\sigma = \frac{\partial^2 \Phi}{\partial p^2} - \frac{1}{p}\frac{\partial \Phi}{\partial p}\left(\frac{R}{C_p} - 1\right)$$

where $\eta = c_v c_p$.

16

SYNOPTIC OBSERVATIONS AND ANALYSIS

16.1 INTRODUCTION

In all fields of science documented observations form a crucial cornerstone. The practice of detailed observation of weather dates from antiquity – yet obtaining a sufficient number of reliable and accurate observations – taxes the organizational and technological skills of even the industrialized nations. Observations of the weather encompass the techniques of classification, of estimation and of measurement. Any objective scheme should ensure that different observers of the same event record the same weather type. Clearly a universal scheme for observing and documenting the weather is required; fortunately such a scheme exists. The remarkable feat of devising this scheme was achieved by meteorologists through the World Meteorological Organization* (formed in 1950) and its predecessor (International Meteorological Organisation*, created in 1873), and is a splendid example of the internationalism, common sense and cooperation that is still traditional amongst meteorologists.

The operational World Weather Watch is a daily example of this cooperation. The World Weather Watch encompasses the Global Observing System which includes over 9000 land stations and 7000 ships. About one in ten of the land stations release weather balloons for upper air observations daily. Ships provide similar reports and are often invaluable when they are in data-sparse areas. Their reports also include sea surface temperature, sea and swell details. Other important data come from satellites, radars, hundreds of drifting ocean buoys, and several thousand commercial aircraft which measure and relay inflight temperature, humidity, wind and weather. Increasingly, national weather services and other agencies are installing automatic weather stations, especially in remote locations. The advantages are that they provide the agency with continuous data of the easily measured quantities such as wind,

---

* The different spellings are indeed correct.

temperature, humidity and pressure (although this last is relatively expensive and therefore sometimes omitted). They can also be preset to send a message when predefined events occur, such as the wind exceeding 30 knots. The disadvantages are that other important parameters of weather type, visibility and cloud are not measured, although the technology to do this is now being developed in several countries. To compensate for this lack, there is often provision for the insertion of additional information by human operators.

16.2 SYNOPTIC OBSERVATIONS AND PLOTTING

The weather classification scheme in use enables an enormous amount of information to be concisely displayed and quickly absorbed by an experienced

STATION MODEL

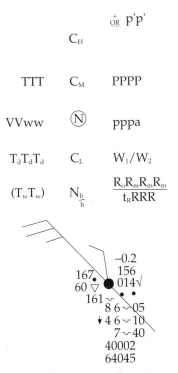

FIGURE 16.1 (a) A model plot of a synoptic observation: TTT = temperature; VVww = visibility and present weather; TdTdTd = dew point temperature; (TwTw) = wet-bulb temperature; C_H = type of high cloud; C_M = type of middle cloud; C_L = type of low cloud; N = number of octas of cloud; $N_{h/h}$ = number of octas of low cloud and height; PPPP = pressure (m.s.l.); $p'p'$ = corrected pressure tendency; pppa = uncorrected pressure tendency; W_1/W_2 = recent weather; R_0 group = rainfall in specified periods. (b) A sample plot showing showers and recent rain with northerly winds. The larger wind plot is the 3000 ft (910 m) balloon wind.

meteorologist of any nationality. Observations from official stations are plotted to display information on weather (phenomena), cloud, wind, temperature, humidity, pressure, visibility, rainfall and other quantities. Figure 16.1 shows a sample plot and a generic plot. Typically, observations are performed every 3 hours. By common usage, the term synoptic observation has come to apply to this type of observation.

Phenomena

The phenomena, or present weather, category has provision for 100 types of weather and these are condensed into a 10 × 10 matrix. The matrix with its 100 symbols is shown in Fig. 16.2. Each symbol and corresponding number has a specific meaning as shown in Table 16.1. Once one knows about 10 symbols, most of the others are easy to remember or guess. Some common ones are ∇ for showers, ❜ for drizzle, ● for rain. Even knowing only a handful of the common symbols, one can quickly scan a group of plotted observations and obtain a reasonable feel for the weather.

Past weather is also recorded. There are two 'past weathers': one for weather during the past hour and one for weather during the past 3 hours. The series of symbols is an abbreviated version of the present weather scheme and has only 10 types. These are shown in Fig. 16.3.

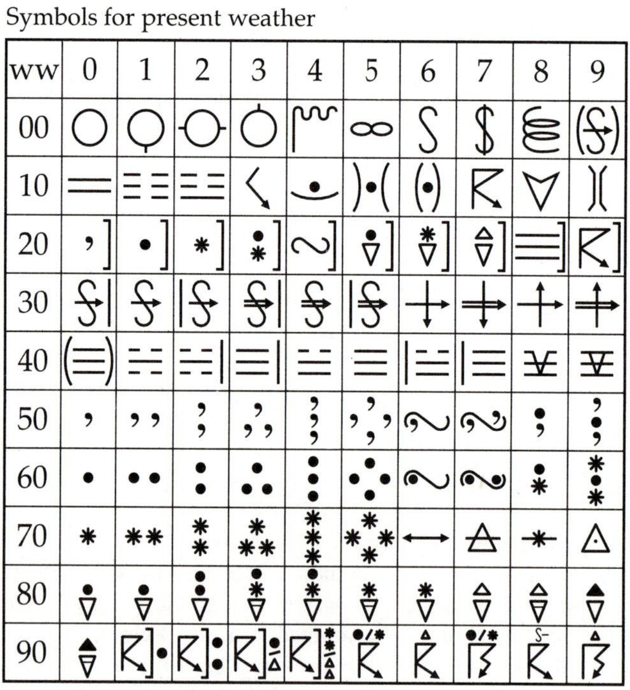

Symbols for present weather

FIGURE 16.2 The symbols for 100 types of weather. Specific meanings are shown in Table 16.1.

Table 16.1 Present weather symbols

No.	Symbol	Explanation
00		Cloud development not observed or not observable
01		Cloud generally dissolving or becoming less developed
02		Cloud on the whole unchanged or sky cloudless
03		Cloud generally forming or developing
04		Smoke originating from bush or industrial fires
05		Haze, very small dry particles, relative humidity below 90%
06		Widespread dust in suspension, not raised at or near the station at the time of observation
07		Raised dust or sand at the time of observation, no well-developed whirls and no duststorm or sandstorm seen
08		Well-developed dust whirls (dust devils) at or near the station at the time of observation or during the past hour
09		Distant duststorm or sandstorm at the time of observation or at the station during the past hour
10		Mist, visibility 1000 m to 10 km inclusive, relative humidity above 90%
11		Shallow fog (below 2 m) in patches

Table 16.1 Continued

No.	Symbol	Explanation
12		Shallow fog (below 2 m) generally continuous
13		Lightning seen, no thunder heard
14		Precipitation not reaching the ground, near to or at a distance from the station
15		Precipitation reaching the ground but not at the station, estimated more than 5 km away
16		Precipitation reaching the ground but not at the station, estimated less than 5 km away
17		Thunderstorm (thunder heard) without precipitation
18		Squall during the past hour or at the time of observation
19		Funnel cloud during the past hour or at the time of observation
20		Drizzle at the station within the past hour, but not at the time of observation
21		Rain at the station within the past hour, but not at the time of observation
22		Snow at the station within the past hour, but not at the time of observation

Table 16.1 Continued

No.	Symbol	Explanation
23		Rain and snow at the station within the past hour, but not at the time of observation
24		Freezing drizzle or rain at the station within the past hour, but not at the time of observation
25		Shower of rain at the station within the past hour, but not at the time of observation
26		Shower of snow at the station within the past hour, but not at the time of observation
27		Shower of hail at the station within the past hour, but not at the time of observation
28		Fog (visibility less than 1000 m) at the station within the past hour, but not at the time of observation
29		Thunderstorm (thunder heard) at the station within the past hour, but not at the time of observation
30		Slight or moderate sand- or duststorm (visibility less than 1000 m but more than 200 m) decreased during the past hour
31		Slight or moderate sand- or duststorm (visibility less than 1000 m but more than 200 m) unchanged during the past hour
32		Slight or moderate sand- or duststorm (visibility less than 1000 m but more than 200 m) began or increased during the past hour

Table 16.1 Continued

No.	Symbol	Explanation
33		Severe sand- or duststorm (visibility less than 200 m) decreased during the past hour
34		Severe sand- or duststorm (visibility less than 200 m) unchanged during the past hour
35		Severe sand- or duststorm (visibility less than 200 m) began or increased during the past hour
36		Drifting snow, slight or moderate below eye level
37		Drifting snow, thick below eye level
38		Blowing snow, slight or moderate below eye level
39		Blowing snow, thick above eye level
40		Fog at a distance, sky visible, at the time of observation, but no fog at the station during the past hour (visibility more than 1000 m)
41		Fog in patches at the time of observation, sky visible (visibility less than 1000 m in patches)
42		Fog in patches at the time of observation became thinner during the past hour, sky visible (visibility less than 1000 m in patches)

Table 16.1 Continued

No.	Symbol	Explanation
43		Fog at the time of observation became thinner during the past hour, sky *invisible* (visibility less than 1000 m)
44		Fog at the time of observation unchanged during the past hour, sky visible (visibility less than 1000 m)
45		Fog at the time of observation unchanged during the past hour, sky *invisible* (visibility less than 1000 m)
46		Fog at the time of observation began or became thicker during the past hour, sky visible (visibility less than 1000 m)
47		Fog at the time of observation began or became thicker during the past hour, sky *invisible* (visibility less than 1000 m)
48		Fog at the time of observation depositing rime, sky visible (visibility less than 1000 m)
49		Fog at the time of observation depositing rime, sky *invisible* (visibility less than 1000 m)
50		Slight intermittent drizzle at the time of observation
51		Slight continuous drizzle at the time of observation
52		Moderate intermittent drizzle at the time of observation
53		Moderate continuous drizzle at the time of observation
54		Heavy intermittent drizzle at the time of observation

Table 16.1 Continued

No.	Symbol	Explanation
55		Heavy continuous drizzle at the time of observation
56		Slight freezing drizzle at the time of observation
57		Moderate or heavy freezing drizzle at the time of observation
58		Slight drizzle with rain at the time of observation
59		Moderate or heavy drizzle with rain at the time of observation
60		Slight intermittent rain at the time of observation
61		Slight continuous rain at the time of observation
62		Moderate intermittent rain at the time of observation
63		Moderate continuous rain at the time of observation
64		Heavy intermittent rain at the time of observation
65		Heavy continuous rain at the time of observation
66		Slight freezing rain at the time of observation
67		Moderate or heavy freezing rain at the time of observation

Table 16.1 Continued

No.	Symbol	Explanation
68	● ✳	Slight rain or drizzle with snow at the time of observation
69	✳ ● ✳	Moderate or heavy rain or drizzle with snow at the time of observation
70	✳	Slight intermittent snow at the time of observation
71	✳ ✳	Slight continuous snow at the time of observation
72	✳ ✳	Moderate intermittent snow at the time of observation
73	✳ ✳ ✳	Moderate continuous snow at the time of observation
74	✳ ✳ ✳	Heavy intermittent snow at the time of observation
75	✳ ✳ ✳ ✳	Heavy continuous snow at the time of observation
76	←→	Ice prisms at the time of observation
77	△	Snow grains at the time of observation
78	—✳—	Isolated star-like crystals at the time of observation
79	△˙	Ice pellets at the time of observation

Table 16.1 Continued

No.	Symbol	Explanation
80		Slight rain shower at the time of observation
81		Moderate or heavy rain shower at the time of observation
82		Violent rain shower at the time of observation
83		Slight rain and snow shower at the time of observation
84		Moderate or heavy rain and snow shower at the time of observation
85		Slight snow shower at the time of observation
86		Moderate or heavy snow shower at the time of observation
87		Slight soft hail shower at the time of observation
88		Moderate or heavy soft hail shower at the time of observation
89		Slight hail shower at the time of observation

Table 16.1 Continued

No.	Symbol	Explanation
90		Moderate or heavy hail shower at the time of observation
91		Thunderstorm in the past hour with slight rain at the time of observation
92		Thunderstorm in the past hour with moderate or heavy rain at the time of observation
93		Thunderstorm in the past hour with slight snow or hail at the time of observation
94		Thunderstorm in the past hour with moderate or heavy snow or hail at the time of observation
95		Slight or moderate thunderstorm with rain and/or snow at the time of observation
96		Slight or moderate thunderstorm with hail at the time of observation
97		Heavy thunderstorm with rain and/or snow at the time of observation
98		Thunderstorm with dust- or sandstorm at the time of observation
99		Heavy thunderstorm with hail at the time of observation

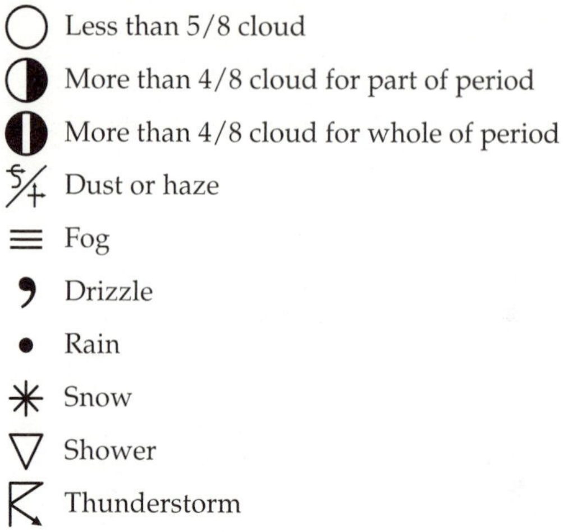

○ Less than 5/8 cloud

◐ More than 4/8 cloud for part of period

◑ More than 4/8 cloud for whole of period

⅘ Dust or haze

≡ Fog

❜ Drizzle

• Rain

✳ Snow

▽ Shower

ᛐ Thunderstorm

FIGURE 16.3 Symbols for past weather.

Cloud

Just as weather has a systemic classification, so too has cloud cover. First, the amount of sky covered is recorded in eighths (usually known as octas or oktas). The number of octas is always rounded up to the next integer excepting the case of rounding up to 8. Completely cloud-free or cloud-covered sky corresponds to 0 or 8 octas respectively. The estimated direction of movement of the lowest cloud is shown by a small arrow plotted adjacent to its symbol.

As with weather symbols, the novice can quickly acquire a reasonable grasp of the cloud through knowing the most common and important ones, for example cumulus, strato-cumulus, cumulo-nimbus. Knowledge of a few Latin words is useful (*cumulus* = heaped; *strata* = layers; *cirrus* = hair; *alto* = middle; *nimbus* = rain). The experience of many perceptive observers has led to the current convention of cloud classification. Clearly, any classification must be based on appearance but the visual features used are those that indicate the physical processes of the cloud's formation. In this sense, there are but two main types of cloud, namely cumuliform and stratiform. Cumuliform comprise cumulus types in which the clouds are usually separated by clear spaces. Stratiform clouds are in sheets or layers covering large areas.

By convention, clouds are also classified into three height ranges (called etages) as low, middle or high as in Table 16.2. Clouds rarely form above the

Table 16.2 Typical cloud base ranges for the low-, middle- and high-level clouds for different regions

Etage	Tropical	Temperate	Polar
High	6–18 km	5–14 km	3–8 km
Middle	2–8 km	2–7 km	2–4 km
Low	up to 2 km	up to 2 km	up to 2 km

FIGURE 16.4 A selection of common cloud types: A, cumulus (humilis) – often known as fair-weather clouds; B, cumulus (congestus) – can produce brief showers; C, cumulus (with vertical development); D, strato-cumulus – low-level layered cloud; E, cumulo-nimbus – will produce showers; F, cumulo-nimbus with anvil – will produce heavy showers, often with thunder and lightning observed; G, cumulo-nimbus (mammatus) – as for F, the downward protuberances are due to strong downdrafts within the cloud; H, arc cloud or squall line – severe winds, heavy showers and

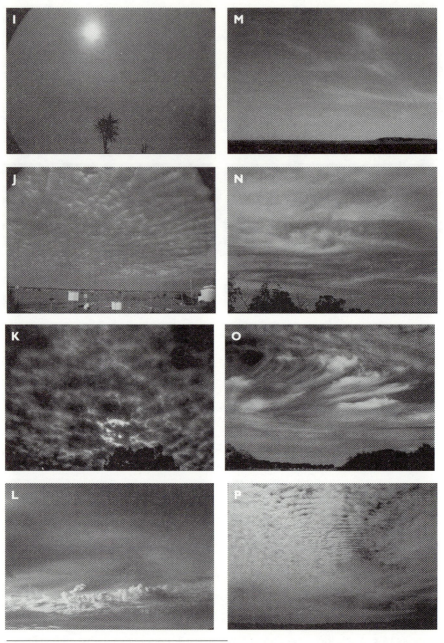

possibly hail expected; I, alto-stratus – middle-level grey, generally uniform, sheet; J, alto-cumulus (undulatus) – rippled elements, generally white; K, alto-cumulus (floccus) – rippled nature but darker; L, alto-cumulus (castellanus) – good indicator of middle-level instability: when observed in the morning, often indicate showers or thunderstorms in the afternoon; M, cirrus (fibratus); N, cirrus dense; O, cirrus uncinus – white tufts or filaments; P, cirro-cumulus – small rippled elements. Clouds A to G are low-level clouds; H to L are middle level and M to P are high level. (Photographs courtesy of Mark Bedson, Cloud World, Ceduna, South Australia 5690.)

troposphere. There are 10 main types of clouds (technically *genera*, the singular being *genus*) and 27 subtypes according to height, shape, colour and weather. Some common cloud types are shown in Fig. 16.4.

Identifying and classifying clouds, and estimating their base heights, is not always easy. Sometimes clouds are intermediate between the standard forms. Trained observers do not merely note the sky at the time of observation but will have been noting the evolution of the clouds as opportunity permits well before the observing time. This continual study will often help classification. Other aids in estimating cloud heights are information from pilots and inspection of an aerological sounding.

The basic symbols of the different cloud types are given in Fig. 16.5. Column 6 (headed 'C') shows cloud type. Column 2 (headed 'N') shows cloud amount in octas. If there is only one type of cloud observed, then column 6 is used, otherwise column 3, 4 or 5 as appropriate (for low, middle and high cloud, respectively).

Visibility

Visibility is a parameter especially important to aviation users. It is a horizontal measure of the transparency of the atmosphere and, as such, should be

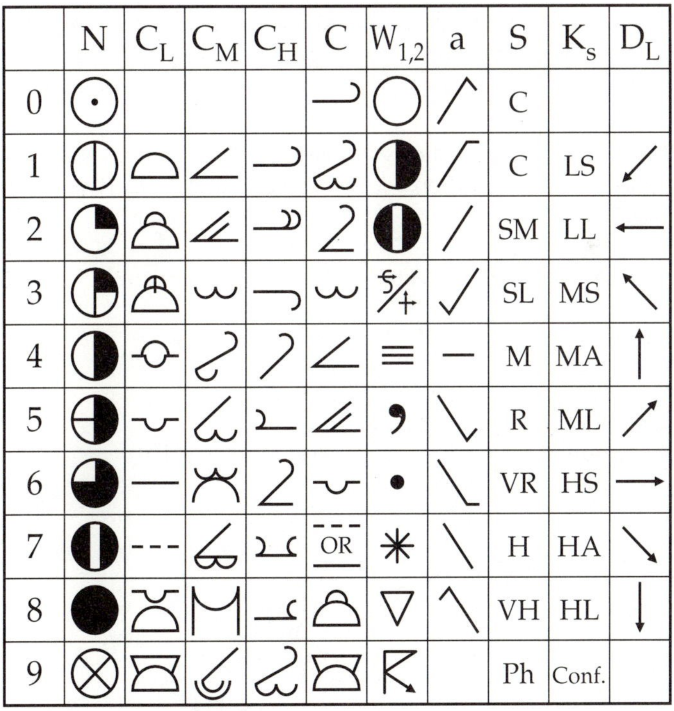

cloud

FIGURE 16.5 Symbols for cloud amounts and types, past weather, sea and swell, and direction of motion of lowest cloud.

unaffected by whether it is daytime or night-time. The visibility is coded as in Table 16.3. Estimates of visibility are arrived at by reference to landmarks obscured by, or discernible through, drizzle, rain, haze, dust, mist, fog, etc.

Rainfall

Rainfall recorded during the previous 3 hours and 6 hours is also plotted.

Wind

Apart from rainfall, none of the preceding categories requires any particular instrument. Wind, too, may be estimated by eye and this visual estimation is performed via the Beaufort scale in Table 16.4. Of course, it is desirable to be more objective and to measure the wind speed and direction with a standard anemometer. Anemometers are located on masts at 10 m above ground level although sometimes 2 m or 3 m elevations are used. The anemometer needs to be located such that the measured wind is unaffected by barriers and obstacles. A good site is one where any obstacle is distant from the anemometer by at least 10 times its height. In practice, it is surprisingly difficult to obtain good measuring sites.

If wind direction is estimated then it is estimated to the nearest of the eight compass points. If direction is measured then it is recorded to the nearest 10° increment from true north. Both speed and direction are the mean values over a 10 minute interval. Direction is taken as direction *from* – a point of confusion for many students of mathematics or physics coming to meteorology for the first

Table 16.3 Left hand portion shows visibility code of *non-aeronautical* stations. Right hand portion shows visibility code for *aeronautical* stations. (Code figures 51 to 55 are not used.)

Criteria: non-aeronautical	Code	Criteria: aeronautical	Code
Object visible at <50 m	90	Object visible at <100 m	00
Object visible at 50 m	91	Object visible at 100 m	01
Object visible at 200 m	92	Object visible at 200 m	02
Object visible at 500 m	93	:	:
Object visible at 1000 m	94	in increments of 100 m to	
Object visible at 2000 m	95	5 km	50
Object visible at 4000 m	96	6 km	56
Object visible at 10 km	97	7 km	57
Object visible at 20 km	98	:	:
Object visible at 50 km	99	in increments of 1 km to	
		30 km	80
		:	:
		in increments of 5 km to	
		70 km	88
		greater than 70 km	89

Table 16.4 Beaufort scheme for estimating wind speed at a height of 10 m on land. A similar scheme is used for estimating speed over the ocean

Specification for estimating wind speed		
Specification	**Description**	**Speed (knots)**
Calm; smoke rises vertically	Calm	0
Smoke drifts; wind vane steady	Light air	2
Wind felt on face; leaves rustle; vane moves	Light breeze	5
Leaves and small twigs moving	Gentle breeze	9
Dust and loose paper raised	Moderate breeze	13
Small leafy trees sway	Fresh breeze	18
Large branches sway; whistling in overhead wires	Strong breeze	24
Whole trees sway; inconvenient to walk	Moderate gale	30
Twigs break off; difficult to walk	Fresh gale	37
Slight structural damage	Strong gale	44
Trees uprooted; considerable structural damage	Storm	52
Widespread damage (rarely experienced on land)	Violent storm	60
Widespread damage (rarely experienced on land)	Hurricane	68

time. The main reason, apart from tradition, is that when using the simplest forecast (based upon advection from upstream) one is concerned with where the air has come *from* rather than where it is going to. (To compound the possible confusion, oceanographers use the opposite convention.)

Another aspect of tradition is in the matter of the units of speed. Traditionally knots have been used, yet the appropriate SI units are metres per second. Fortunately, since $1\,\mathrm{m\,s^{-1}} \approx 2$ knots, the same plotting scheme of barbed, or feathered, arrows is applicable. For example, a full feather on the arrow shaft equals 10 knots or $5\,\mathrm{m\,s^{-1}}$. Feathers are to the left in the northern hemisphere and to the right in the south, the idea being that if the wind were geostrophic, the feathers would point to the low pressure. If the maximum gust is significantly above the mean it is sometimes included near the wind tail as '<28', for example, to indicate a gust to 28 knots. Calm or light variable winds are indicated by a solid circular line or dashed circular line respectively.

Temperature and humidity

Temperature is obtained to the nearest tenth of a degree from a thermometer housed in a Stevenson screen and plotted omitting the decimal point. In some offices, the plotted quantity is rounded to the whole degree. Similarly dew point temperature is plotted a little below the dry-bulb temperature. Relative humidity is not normally plotted because this is not a conservative quantity whereas dew point temperature is approximately so (see problem 2).

As a general rule, no data should ever be accepted unquestioningly. Of all the standard meteorological measurements, the quantity which seems to be the least reliable is humidity. In particular, the accuracy and reliability of dew point temperatures suffer from incorrect wet-bulb temperature readings due to poorly maintained equipment, and faults such as dried or dirty muslin coverings on the thermometer, or algae in the water reservoir. These faults usually lead to an overestimation of the dew point temperature.

Pressure and its tendency

In terms of the standard analysis of the observations, the most important quantity is pressure. The barometer provides station-level pressure (SLP). To this value must be added a correction to give what the pressure would be at mean sea level (m.s.l.) – this is called reducing to sea level. This correction adds a factor for the station's altitude on the assumption of a mean virtual temperature in a hypothetical column of air from m.s.l. to the barometer altitude. The mean temperature of the column is assumed to be determined by the temperature at the station and a standard lapse rate of $6.5°C\,km^{-1}$ over the height of the column. In practice, there are several variations on this procedure. For example, many countries use for the station temperature an average of the current and the 12 hour prior temperature while some use the climatological mean. The correction is calculated through the hypsometric equation (5.3) which is recast as

$$p_{msl} = p_{sl} \exp\left(\frac{gH}{R_d T_m}\right) \qquad (16.1)$$

where p_{msl} and p_{sl} refer to the reduced pressure and the station-level pressure respectively. H is the station height and T_m is the mean temperature of the column.

The observed tendency of any quantity is usually defined as the latest reading minus the previous reading. For pressure, two SLP tendencies over the past 3 hours are given. First, an *uncorrected* tendency in tenths of hectopascals is given by selecting from a choice of eight symbols to describe best the barograph trace. If no barograph is available then a simple 2, 4 or 7 is used to indicate whether the SLP has risen, showed no net change or fallen, respectively. Secondly, the *corrected* tendency is plotted. This correction compensates the observed 3 hourly change in SLP by subtracting out the expected 'tidal' or mean variation for that time of day and year. This corrected tendency is the one that appears in the pressure tendency equation in section 9.4 as $\partial p/\partial t$ although conversion from hPa per 3 hours to SI units is required. The pressure tendency is also known as the isallobaric component (see Chapter 9). The isallobaric component may be determined over other periods, for example 24 hours, in order to analyse a daily rise/fall pattern.

The approach of a low-pressure system will often be heralded by rapidly falling pressure. Usually, isallobaric components will be large when synoptic systems are approaching quickly or developing and therefore the isallobaric component is carefully monitored by forecasters. However, there are other small-scale effects which can produce large components and these need to be borne in mind. Examples are thunderstorm outflows, and a 'pumping' or chimney effect of strong winds on barometers in lighthouses – this is of significance because a lighthouse is often the closest observing site to approaching weather systems. An example of such pumping during strong prefrontal northerly flow (southern hemisphere) at a lighthouse is shown in Fig. 16.6.

Customizing

Tendencies may be calculated for other parameters also, especially temperature and dew point. These provide additional information useful for analysis in some circumstances. Deriving and plotting 24 hour tendencies of temperature and dew point temperature at stations either side of a quasi-stationary front will often provide the first objective sign, or confirmation, of the formation and deepening of small-scale wave lows on the front.

All forecasting centres have their own styles. Commonly used colour associations are yellow for fog, green for rainfall, red for thunderstorms, red also for falling pressure and rising temperatures with blue for rising pressure and falling temperatures. Thus, for example, a red-blue boundary will emphasise a frontal zone. Other derived quantities such as dew point depression, fire danger index, soil dryness index and other thermodynamic parameters and indices are often plotted, depending upon the season and the circumstances.

Although the plotting of the observations synoptically (i.e. in a spatial pattern for a 'snapshot' in time) is the norm, one may also produce a time series for one or more stations (*see* Fig. 16.7). Commonly, a synoptic chart will cover an area with dimensions of 2000–4000 km. Adequately covering all the stations and retaining legibility requires charts of physical size of about a square metre, and

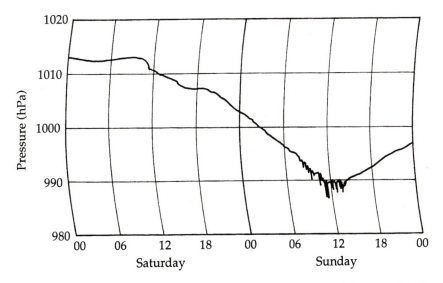

FIGURE 16.6 Barometer pumping of up to 3 hPa between 8 a.m. and 2 p.m. on Sunday during a frontal passage at a lighthouse.

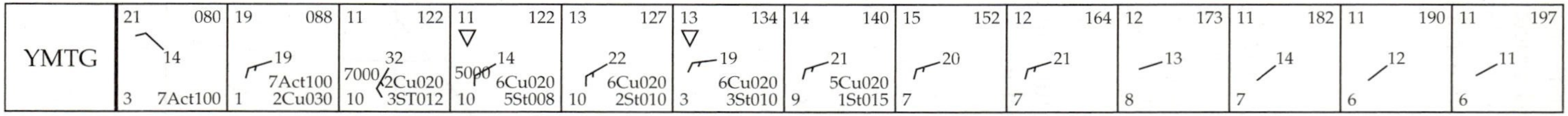

YMTG	21	080	19	088	11	122	11	122	13	127	13	134	14	140	15	152	12	164	12	173	11	182	11	190	11	197
		14		19		32		14		22		19		21		20		21		13		14		12		11
		7Act100		7Act100 2Cu030		7000 2Cu020 3ST012		500 6Cu020 5st008		6Cu020 2St010		6Cu020 3St010		5Cu020 1St015											
	3		1		10		10		10		3		9		7		7		8		7		6		6

FIGURE 16.7 Time series of airport weather observations (abbrebiated synoptic observations). Note the poor visibility and low cloud at 0200 and 0300 UTC.

unfortunately beyond the scope of this book to reproduce. However, an increasing number of charts are becoming available on the World Wide Web.

The main purpose in plotting the observations is to analyse them, and by doing so, to make judgements about physical causes of the current and past weather – and thence to make further judgements about future weather.

16.3 ANALYSIS METHODS

In polar and temperate regions, weather is mostly associated with the features on an m.s.l. pressure analysis. The charts depicting these features are usually referred to as synoptic. Hitherto we have rather blithely glossed over any definition of *synoptic*. A dictionary will show that the original and literal meaning is 'at that time'. So-called synoptic charts generally cover geographic dimensions of 3000 km or more. Common usage has produced the term synoptic for the features typically seen on these charts, and, by extension, these synoptic features have a synoptic scale of the order of 1000 km.

The horizontal scale of synoptic features is typically about 2000 km and the vertical scale is about 10 km (altitude of the tropopause): the typical wind speeds are respectively 20 m s^{-1} and 10 cm s^{-1}. Thus, the time taken for an air parcel to traverse a stationary system is roughly equal to the time taken for it to travel through its vertical extent, that is 1 to 2 days. Denoting the vertical and horizontal length and speed scales respectively as H, L, W and V, then a synoptic scale ratio, S, may be defined as

$$S = \frac{HV}{LW} \qquad (16.2)$$

If $S \sim 1$, then the scales are said to be matched. This simple concept of scale matching can be quite powerful and will be encountered again in a more sophisticated form in relation to tropical cyclone genesis in Section 18.2.

In the first half of this chapter we saw how individual observations were acquired and plotted for display. To synthesize and visualize the patterns inherent in the information in the aggregations of these observations it is necessary to analyse them. There are many types of analyses. The most well known is that of the m.s.l. pressure analysis, often simply called the surface chart, the weather map, or the m.s.l. anal. In general, the most important single parameter analysed on these charts is the horizontal wind field. Modern meteorological analysis is more concerned with air streams (dynamic) rather than air masses (static) and this fundamental perspective leads to a focus on the wind field.

Analysis is a procedure similar to diagnosis. Analysis is the drawing of fields of wind flow, pressure, temperature, humidity and so on, with the purpose of showing what is happening and what has happened; diagnosis is the how and why. Prognosis is then the extrapolation of the fields to the future. Prior to analysis, the surface observations are plotted on a suitable map at their corresponding locations. Drawing of the desired fields is accomplished in two different ways: objectively by computer according to algorithms of varying sophistication; and subjectively. Strictly speaking, plotting of the observations on paper or a computer screen is required only for subjective analysis.

16.3.1 Objective analysis

The objective analysis schemes range from simple contouring by (a) linear interpolation, (b) spline interpolation, (c) least-squares fitting, (d) Kriging methods and (e) Bezier curves, to interpolation schemes by (f) Cressman and (g) Barnes, to complex methods of (h) optimum interpolation which utilize the statistical behaviour of the observations and a first-guess field, (i) initialization and (j) assimilation. Detailed descriptions are beyond the scope of this text, but numerical analysis texts often cover (a) to (e). As a digression, the authors' experience is that isopleths drawn using the Bezier curves, available in many software drawing packages, often approximate closely the forecasters' freehand isopleths.

Objective analysis is often less useful in data-sparse areas and in complex areas. Unfortunately data-sparse areas will continue to exist for a long time yet, and the complex areas are usually those areas which are most interesting (i.e. where the weather is happening). A sophisticated analysis method (a) incorporates neighbouring observations with different error characteristics in an optimum interpolation, (b) includes non-standard-level information, and (c) calls upon information from previous charts to form a consistent analysis in four dimensions (4D, i.e. in x, y, z and time). This aspect of consistency applies to both objective and subjective methods, and means that all data must be scrutinized and checked for quality. In the objective method, the underlying equations such as the hydrostatic equation and the thermal wind equation are used as checks at each grid point in three-dimensional space. In the subjective method, the analyst must have in mind suitable models, of at least an empirical or idealized physical type, that are used as guidance. For example, at higher latitudes the winds will tend to be parallel to isobars and the speeds will be related to the pressure gradient. There are many exceptions to simple rules of this type, however.

Computer models which predict the state of the atmosphere must start of course with initial conditions. The accuracy of these initial conditions is extremely important in that it is now well established that small differences in the initial conditions can sometimes lead to large differences in model prognoses at a later time. The treatment of the initial observational data and their analysis is an important and non-trivial aspect known as initialization. The number of parameters to be determined explicitly for starting a global numerical forecast model is in the range of 10^6 to 10^7, whereas the number of individual scalar observations available for a 24 hour period is about 10^5. The effect of this discrepancy is partially countered by the processes of initilization and assimilation. Initialization typically involves running the prognostic equations back and forth to achieve an optimization between fitting the observed data (at time $= 0$) and physical consistency. A more complex process known as assimilation is used to incorporate observations received late and those performed at non-synoptic times, or those that have required correction. Sequential assimilation entails running the numerical model over the period of assimilation (which ends at time = 0). Whenever the model time reaches the time of an observation, that observation is incorporated by using its data in lieu of the model's estimate. Variational assimilation in 4D, which is the more general problem, attempts to find a model 'trajectory' which best fits the observations from the assimilation period *and* the forecast field valid at time = 0. The inclusion of the forecast field as a set of pseudo-observations has the positive effects of enforcing the dynamics of the primitive equations in the fields and of retaining information from time prior to the assimilation period. The retention of previous information has a further benefit: that is, the prior information from areas with good data coverage is, in effect, advected into regions where data coverage is poor. The technique of assimilation provides the means by which satellite data and data from other sources can be ingested and its inherent information fully utilized. Another form of assimilation is the technique of nudging. In this process the model is initialized. Then, the vertical motion field, which is susceptible to error as explained in earlier chapters, is derived from satellite cloud imagery via heating considerations. This upgraded vertical motion field is then inserted into the model. The computational process of assimilation in 4D is extremely complex and is, at the time of writing, yet to be fully developed for operational use. It is worth bearing in mind that, ultimately, objective analyses are subjective in the sense that someone had to choose the analysis scheme subjectively.

16.3.2 Subjective analysis

Subjective analysis is more of an art than a science, yet despite the subjectivity, experienced analysts given the identical unanalysed charts will produce very similar analysed charts. This is because they will draw their contours not only by the values of the relatively few data points but by using accepted physical and descriptive models of the atmosphere. Compared with powerful computers with modern pattern recognition software, humans are extremely good at visually identifying and discerning coherent features even as those features move and evolve. Subjective analysis relies upon the individual consciously or unconsciously absorbing a vast amount of data (some of which may be inaccurate or even misleadingly wrong) and synthesizing, or creating, a representation on the chart. A good analysis is really a good diagnosis – a clear understanding of what is happening and why. This is a major step in deciding what will happen later. Experience in subjective analysis is priceless. The eminent Jule Charney is quoted thus:

> I think a person who has made subjective analyses of weather maps has a deeper appreciation of how inadequate many objective analysis schemes are, and if you simply accepted the machine product as reality, it would be very dangerous. And I think that the tendency to rely too heavily on the machine has occasioned a lot of errors in operational work, and not only in operational work, but in research, synoptic research. (Platzman, 1990, p. 462).

A management text by Reddin draws a similar conclusion regarding the value of such experience. He asserts that 'The ideal Chief Executive Officer would be [among other things] ... a student of weather maps ... [because] ... studying weather maps would teach decision-making under uncertainty' (Reddin, 1985, p. 41).

To obtain the best of both worlds, interactive analysis, where the human analyst can re-arrange and improve the computer's effort, is now a common practice in national meteorological centres. One method to achieve this has created the hybrid art form of 'bogussing' – inserting synthetic observations to force the computer analysis into a 'better' pattern. The operator inserts the synthetic data, which have been estimated subjectively or obtained from other sources such as satellite imagery, usually at a set of predefined grid points over data-sparse areas and at the synoptic features of most interest. From satellite imagery, for example an analyst may note a tropical cyclone in the data-sparse areas of the tropical oceans. The cyclone may well not show up at all on an objective analysis. So the analyst may insert a complete preprogrammed packet of 'bogus' or synthetic observations to simulate a tropical cyclone. A similar concept is sometimes used for mid-latitude cyclones. For example, Japanese analysts use the diagrammatic relationship (in Fig. 16.8) between satellite imagery of cloud associated with a depression and the pressure field (Ryoji Kumabe, personal communication). This classification follows the scheme of Smigielski and Mogil (1992) developed for Atlantic lows but is 'tuned' for the area around Japan.

One intriguing aspect of analysis is that of aesthetics. Forecasters sometimes speak of a chart as 'passing the look test', and regard an analysis as incorrect if it is 'ugly'. Jerome Namias gives an instance of where his wife, an artist, completed one of his sea surface temperature analyses which he had left unfinished awaiting more data. When challenged to justify her analysis, she remarked that her interpretation was justified because it was artistically pleasing. To Namias' chagrin, when the remainder of the data was later plotted, his wife turned out to be correct. A quote from a UK Meteorological Office handbook also emphasizes this aesthetic side:

> Nature is at heart an artist and the smooth, functional lines and colours of the synoptic chart show ample evidence of this truth. ... The final chart should have artistic beauty and give pleasure to the eye of the viewer. ... the chart should have some of the beauty of a painting or other work of art. ... A synoptic chart so drawn will convey at one glance something of the working of the atmosphere to the viewer'. (UKMO, 1964, p. 19).

16.3.3 Streamlines

Analysis in tropical areas is qualitatively different mainly because of the absence of fronts and the smallness of the Coriolis force. Within $20°$ of the equator, the geostrophic 'rule' is of little use and the isobaric spacing is usually too open to provide meaningful analyses of the wind field. To overcome this difficulty, streamline analysis is often used.

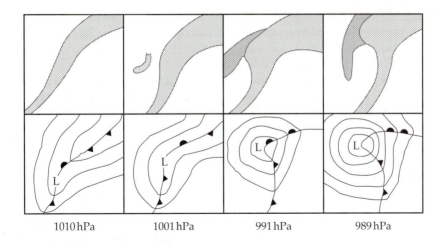

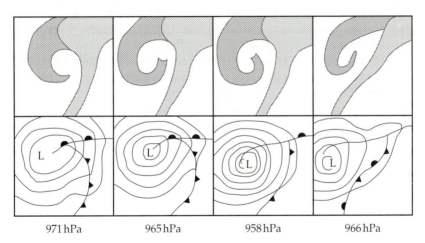

FIGURE 16.8 Schematic relation between cloud imagery and m.s.l. pressure pattern for cyclones around Japan.

Streamline analysis entails drawing a line on a chart to represent the wind direction at that time. The only constraint is that the wind direction should be tangential to the streamline. Spacing is simply a matter of convenience. Speed is analysed separately (although usually on the same chart) by isotach analysis. Note that a streamline is not the same as a trajectory; however, the two are identical only in steady-state conditions. Streamline analysis can be used over areas poleward of 20° and sometimes forecasters perform hybrid blends of streamline and isobaric analyses on working charts, especially over data-sparse areas.

16.3.4 Trends

The m.s.l. pressure analysis remains the most complex chart of all. Most meteorologists consider that the most important routine charts (either isobaric or streamline) are the m.s.l. analysis, with the thickness (or 500 hPa) analysis and jet stream level (say 250 hPa), and this is likely to continue to be the case for some time. Forecasters will likely focus on more frequent analyses of a subsynoptic or regional scale with the larger scale being automated. These local area analyses will be integrated with a vast amount of satellite and radar imagery leading to a whole new set of subjective skills of an empirical and experiential nature in the identification and interpretation of meteorological features and the prediction of their evolution. However, these interpretive skills will not be confined to local areas. Over data-sparse areas, cloud patterns from satellite imagery can be classified and assigned to idealized pressure patterns.

With the advance of assimilation and pattern recognition techniques for computer analysis and the integration of a vast amount of spatial data, such as TOVS (temperature soundings derived from satellite data), cloud-drift winds, sea surface winds and, potentially, even m.s.l. pressure itself, from satellite and radar, objective analyses produced by computer will undoubtedly improve and match subjective analysis. A further advantage of objective analysis is that additional fields of derived parameters, such as vorticity, potential temperature and Q vectors, can be produced when required. Historically, there has been recurring support for analysing, on a routine basis, the flow on isentropic surfaces rather than, or as well as, isobaric surfaces. The main advantage of these charts is that it is usually very easy to diagnose the vertical motion. These could become the conventional charts of the future.

Much of the foregoing has implicitly demonstrated the ever increasing reliance on computers. It is instructive to look at the growth in computing power used by the European Centre for Medium Range Weather Forecasting. Over the past two decades the power, as measured in gigaflops, of the Centre's main computer has grown exponentially (the plot being roughly a straight line on the log–linear graph in Figure 16.9).

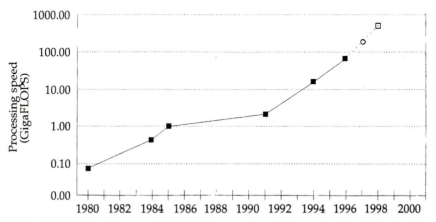

FIGURE 16.9 Plot of power of ECMWF's computer for numerical modelling against time. The approximate straight line nature of the plot on log–linear scales corresponds to exponential growth. Open square indicates projected speed and open circle indicates world's fastest computer as at May 1997.

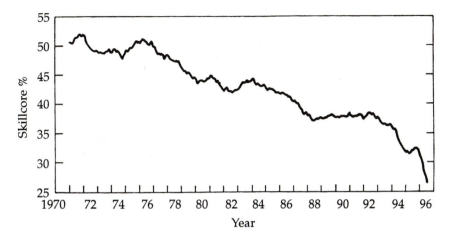

FIGURE 16.10 Plot of S1 skill score of Australian Bureau of Meteorology's model. The downward trend indicates a steadily improving model.

Most national weather services show similar growth in computer power used in numerical modelling. One might ask: how much have the prognoses improved? In Figure 16.10, the skill score of the Australian Bureau of Meteorology's model is plotted against time. The particular score used here is S1 which is essentially a correlation of the pressure gradient at all grid points: the correlation is then subtracted from one and converted to a percentage. So improvements in the model are evident from the marked downward trend in the SI skill score. Many other objectively measured improvements in analysis and forecasting are published in annual reports from national weather services.

16.4 PROBLEMS

1. Air at a temperature of 20°C and pressure of 1000 hPa has a wet-bulb temperature of 10°C. What is the dew point, the relative humidity and the mixing ratio? [Hint: Review Section 4.21.]
2. The concept of aesthetic appeal is an intriguing one. Do you agree with the UK Met Office quote? Or is it simply a case of beauty being in the eye of the beholder?
3. A station at 300 m altitude reports a station-level pressure of 980 hPa. The surface (virtual) temperature for use as the upper 'anchor' point in the fictitious column to sea level is 0°C. What is the pressure reduced to sea level? Repeat for a surface temperature of 5°C.
4. Inspect your local weather office site. Does it have good siting as far as wind is concerned?
5. For 1 week collect the daily m.s.l. analyses in your local newspaper or from some other source. Identify observations which tend to obey geostrophy, and those which tend not to. Offer some plausible reason for the difference.
6. Use the daily charts from problem 5 to estimate the speeds of fronts, lows and highs. Which tend to move faster?

17

SIMPLE SYNOPTIC MODELS

17.1 INTRODUCTION

From a pragmatic perspective, synoptic features are important because weather at a site usually results from the movement of the features across the site. The synoptic features themselves are continuously changing by modifying, intensifying, interacting, weakening, decaying and reforming. Understanding the evolution of the features and the associated weather begins with a knowledge of some descriptive and conceptual models of synoptic features.

Compared with daily charts, the climatological charts which show analyses of the mean pressure fields (monthly or long term) are relatively featureless. It is beyond the scope of this book to present a comprehensive set of climatological charts and it is expected that the reader is familiar with, or, at least, has access to, suitable reference material. Only through experience or systematic study of many synoptic systems and features can one appreciate the similarities and diversities of the individual features. Operational forecasters acquire an experience and knowledge of the typical features, their most common tracks, the most likely areas of intensification or decay, the typical speeds of different types of systems for the different seasons and even for the time of day. This type of knowledge is usually documented and available in various formats such as internal technical publications from the national weather services.

17.2 SOME COMMON SYNOPTIC PATTERNS

Analysed m.s.l. charts show ubiquitous features such as highs, ridges, lows, troughs, cols, 'cut offs', secondary lows and dumb-belling depressions. These features are shown schematically in Fig 17.1. Note that northern and southern hemisphere systems are mirror images (reflections in a mirror aligned with the equator). Fronts are described in a later section.

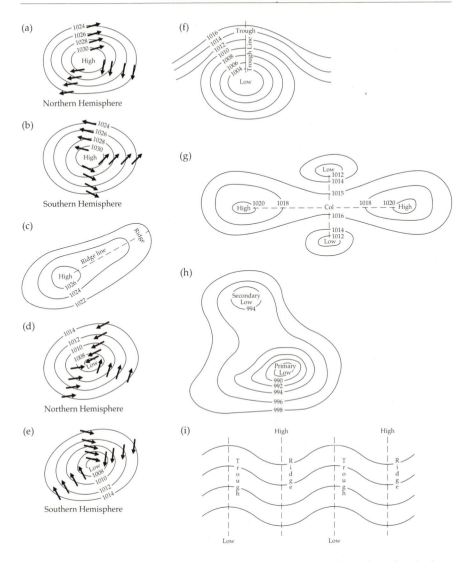

FIGURE 17.1 (a) Northern hemisphere high or anticyclone; (b) southern hemisphere high or anticyclone; (c) ridge of high pressure – either hemisphere; (d) northern hemisphere low or cyclone; (e) southern hemisphere low or cyclone; (f) trough of low pressure – either hemisphere; (g) a col (or neutral point) – either hemisphere; (h) primary and secondary lows – either hemisphere; (i) waves – either hemisphere.

17.3 WEATHER ASSOCIATED WITH SYNOPTIC SYSTEMS

This section applies to tropical as well as extra-tropical systems. The weather associated with fronts is discussed later.

With anticyclones, ridges and zones of surface divergence, the stability is greatest near the centre. If the air is dry, cloudless conditions with light variable winds are likely with dew or frost at night. If the air is moist, mists and fogs may occur at night; during the day, strato-cumulus is likely with drizzle being possible if the cloud depth is sufficient. In cities, under an overcast sky of such strato-cumulus, air pollution is trapped and leads to 'anticyclonic gloom'. Away from the centres of the anticyclones, the wind is stronger and the air less stable. Subsidence aloft will be weaker. The actual weather depends on the air stream characteristics of the stability and moisture content of the air at lower levels, which in turn is modified by the surface itself.

With depressions (non-frontal), troughs and zones of surface convergence, widespread ascent occurs and the weather depends upon the characteristics of the stream. With sufficient moisture, widespread rain will occur. Some depressions are simply the result of intense surface heating and are called heat lows. These systems occur inland and are little more than wind circulations since the lack of moisture precludes significant cloud formation.

A cold stream is one in which the air mass is colder at low levels than the surface over which it is moving. It is therefore heated from below and becomes less stable. The depth and degree of instability depends on the amount of warming and the initial stability. The characteristic cloud of a cold stream is cumulus which may develop into cumulo-nimbus. Showers, hail and thunderstorms may occur.

A warm stream develops when warm air flows over a colder surface. This has the effect of stabilizing the air. If the air is moving over the ocean, then the lower layers may become saturated with fog or low stratus subsequently forming.

17.4 DEFINITION OF A FRONT

Systematic analyses became possible in the late nineteenth century with the widespread adoption of the telegraph and dissemination of observations. It was found possible to track the passage of storms and weather events. However, not for another half-century did the concept of fronts and the life cycles of frontal cyclones or depressions arise. The influential Norwegian meteorologist, Vilhelm Bjerknes, remarked that 'During 50 years meteorologists all over the world had looked at weather maps without discovering their most important features'. A front, or frontal zone, has no universally accepted and clear-cut definition. *The basic concept of a front is that of a boundary.* Often this boundary is several kilometres across and so the term frontal zone is more appropriate. A frontal zone separates air masses of different densities, temperature, humidity or some other physical property. The derived parameter of vorticity generally shows a high degree of consistency and is an effective frontal marker.

Fronts are marked at their surface position and described as cold or warm (or stationary) according to the direction of advance. For example, with a cold front, cold air replaces warm air. In the vertical, fronts slope upwards at very small angles – typically 1 : 70 for cold fronts and 1 : 250 for warm fronts. Vertical profiles of cold, warm and stationary fronts are shown schematically in Fig. 17.2.

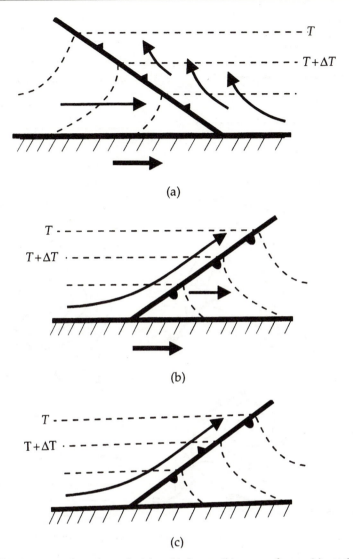

FIGURE 17.2 Cross-section through (a) cold front, (b) warm front, (c) stationary front. Air motion is indicated by arrows; frontal motion by thick arrows. Dashed lines are isotherms.

The most important front is the polar front which separates polar air from tropical air. In an idealized picture, the polar front circles the earth in the middle latitudes in both hemispheres. The warm air on the equatorward side is termed 'tropical' and that poleward 'polar'. Typical temperature gradients are 5°C per 100 km. In some regions along the polar front, colder, denser polar air advances equatorward. This portion of the polar front is known as a cold front; similarly, that portion moving poleward is known as a warm front.

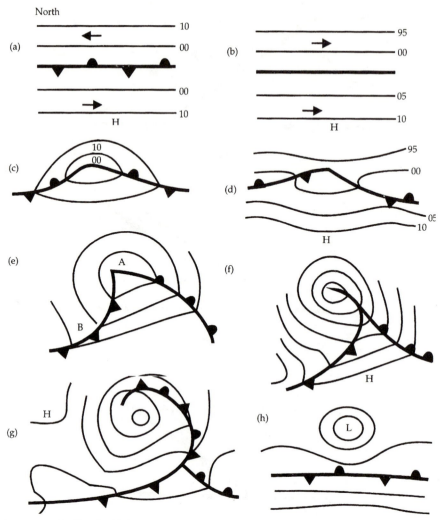

FIGURE 17.3 Classical life cycle of wave cyclone – northern and southern hemisphere. See text for details.

17.5 EVOLUTION OF A WAVE DEPRESSION

The evolution of a wave depression, or polar front cyclone, from a quasi-stationary polar front is shown in Fig. 17.3. Two possible initial states, (a) one with a coincident trough of low pressure and (b) one without, may undergo perturbation and evolve to stages (c) and (d) respectively. Both of these evolve to (e) and thence to (f), (g) and then frontolysis (decay) at (h). During this sequence, the cyclone propagates along the front with the speed and direction of the wind in the warm sector. Rainfall and cloud varies from system to system depending on stability and moisture. At (e), (f), (g) is shown the occluded front. This latter is commonly thought of as the cold front

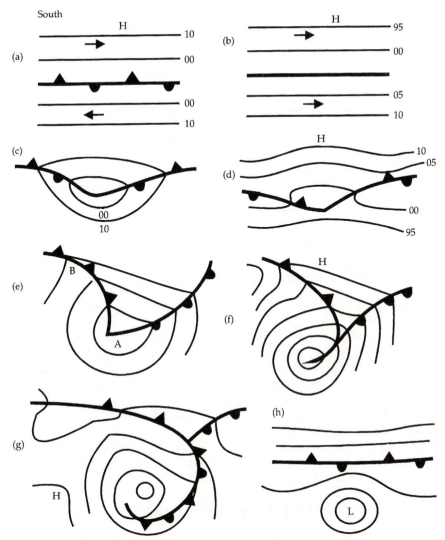

FIGURE 17.3 Continued.

catching up with the warm front, in which case either a cold frontal occlusion or a warm frontal occlusion occurs (*see* Fig. 17.4). However, in reality it is more frequently the separation, and subsequent deepening, of the surface lows from the 'V' junction of the cold and warm fronts poleward into the cold air with a consequent trough line.

The initial perturbation may be due to topography, surface temperature contrast (land and sea, for example), divergence aloft as an upper trough approaches, horizontal wind shear, or the approach of another synoptic feature. Once a front is perturbed, the atmosphere is baroclinic around the perturbation and self-development occurs through to maturity (g) when the low

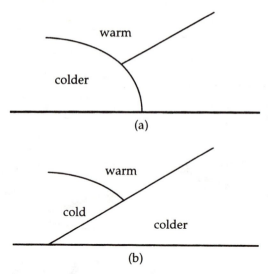

FIGURE 17.4 Model of frontal occlusion: (a) cold; (b) warm.

tends gradually to fill in as the baroclinicity is destroyed. It then frontolyses (decays) as at (h).

During this growth phase, the cold front to the equator–west is strengthened by advection of cold air moving equatorward on the western flank. Cold air increases the surface pressure and the high-pressure cell, labelled H_2 in Fig. 17.5, is induced to form on the poleward side with the front becoming quasi-stationary just equatorward of the high. Yet another wave may form at this point in this manner. If so, a train of waves, or a cyclone family, forms. Each wave brings polar air further equatorward with a polar outbreak occurring in the western flank of the last member. Simultaneously warm air is carried poleward ahead of the subtropical high H. Of course, the above description is highly idealized and each frontal system differs from previous ones.

17.6 FRONTAL THEORY

Frontal theory uses the concept of flow relative to the front. Hoskins and Bretherton (1972) developed such a model (*see* Fig. 17.6) of which the following account follows the exposition of Bennetts *et al.* (1988).

Consider a dry, adiabatic frictionless flow of (U, V) with a straight front moving steadily eastwards with the geostrophic wind speed, u_g, as in Fig. 17.6. The wind relative to the front is therefore $(U - u_g, V)$ or (u, v). The relevant equations in the isobaric system, which the reader should recognize as similar to those of earlier chapters, are

$$v = f^{-1}\partial\phi/\partial x \tag{17.1}$$

$$\mathrm{d}v/\mathrm{d}t + fu = 0 \tag{17.2}$$

$$\partial\phi/\partial z = g\theta/\theta_0 \tag{17.3}$$

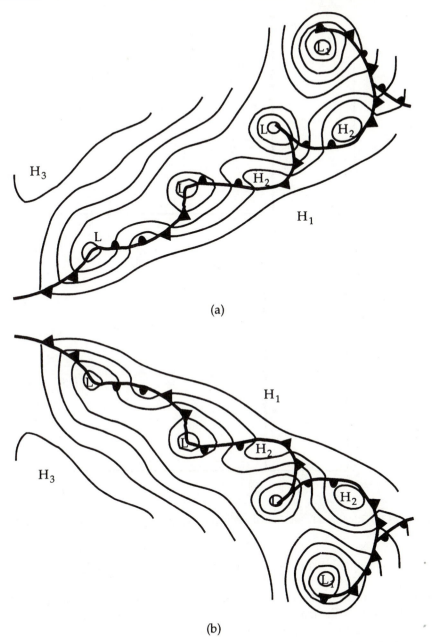

FIGURE 17.5 The wave cyclone family: (a) northern and (b) southern hemisphere.

$$d\theta/dt = 0 \qquad (17.4)$$

$$\partial u/\partial x + \partial w/\partial z = 0 \qquad (17.5)$$

where θ_0 is a reference temperature and z is the height in an isentropic atmosphere. The equations here may be regarded as reformulations of the

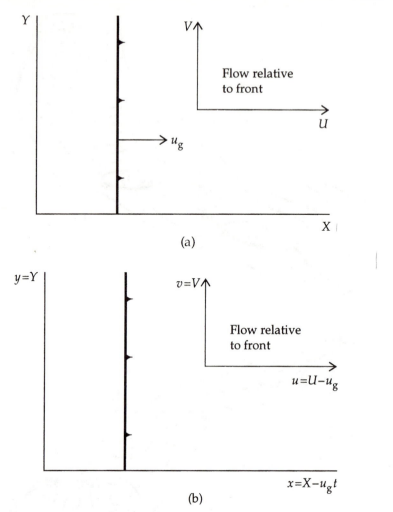

FIGURE 17.6 Idealized front: (a) with flow relative to the ground; (b) with flow relative to the front.

momentum equation, the hydrostatic equation and the continuity equation in an isentropic coordinate sysytem. Equation (17.1) says that the component of wind parallel to the front is always in geostrophic balance. Any ageostrophic motion is therefore across the front. Elimination of ϕ from equations (17.1) and (17.3) gives

$$f \frac{\partial v}{\partial z} = \frac{g}{\theta_0} \frac{\partial \theta}{\partial x} \qquad (17.6)$$

which is actually the thermal wind equation recast. The solution to these equations for a cold front and a warm front is schematically shown by the streamline flow in Fig. 17.7. Note the up motion in the warm air, down motion in the cold air and the cross-frontal flow. This cross-frontal flow is not completely

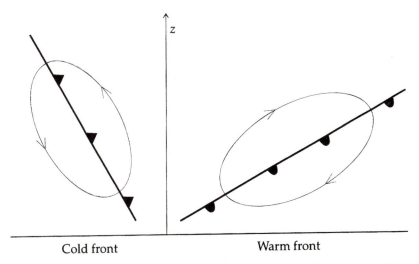

Cold front   Warm front

FIGURE 17.7 Naïve model of cross-frontal ageostrophic motion about cold fronts and warm fronts. The flow through the frontal boundaries is not realistic.

realistic as we have already said that fronts tend to act as material boundaries, and will need to be rectified by some other factor.

Several flow patterns have the property of increasing existing temperature gradients. Three types are shown in Fig. 17.8. These patterns act to intensify the thermal gradient. This intensification via positive feedback occurs thus:

- tightening of thermal gradient at, for example, A and B, as located in Fig. 17.3(e), due to shear and deformation patterns in the low-level flow ($\partial\theta/\partial x$ increases);
- since the thermal wind is in balance along the front, then by equation (17.6) v increases;
- since v increases with time, then by equation (17.2) u increases;
- this increase in cross-frontal ageostrophic motion acts to increase the thermal gradient;
- and so on ... but eventually limited by friction, and ultimately by the synoptic-scale change occurring in the thermal pattern.

In Fig. 17.7 there is cross-frontal flow, but in reality prefrontal jets form as a consequence of equation (17.6). Because air can enter or exit via the jets, then the need for cross-frontal flow is obviated: this is necessary in the sense that otherwise the front would be destroyed. The air entering these jets ascends – and if the air is moist enough, cloud and rain form. Two types of ascending *conveyor belts* of air have been described in models by Browning (1985): namely, a warm conveyor belt and a cold conveyor belt. The cold conveyor belt tends to come in parallel to the warm front and ascend to mid-levels while the warm conveyor belt tends to parallel the cold front prior to its anticyclonic exit at upper levels in the vicinity of the surface position of the warm front (*see* Fig. 17.9). The warm conveyor belt is very commonly seen on satellite pictures. In Australia, it

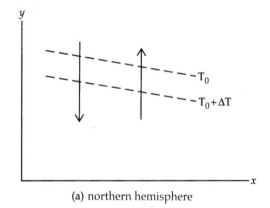

(a) northern hemisphere

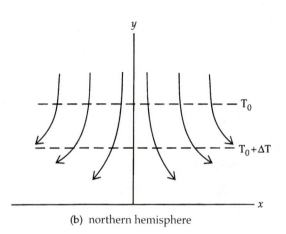

(b) northern hemisphere

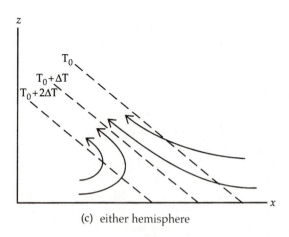

(c) either hemisphere

FIGURE 17.8 Synoptic-scale flow patterns which can intensify horizontal temperature gradient: (a) horizontal shear; (b) horizontal deformation; (c) vertical deformation.

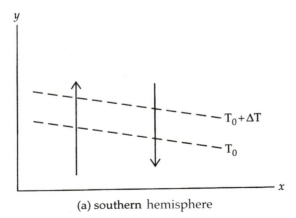

(a) southern hemisphere

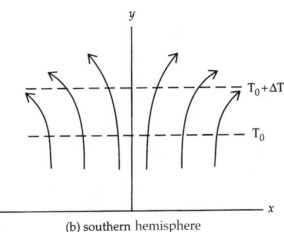

(b) southern hemisphere

FIGURE 17.8 Continued.

may sweep for thousands of kilometres diagonally across the continent and is known as the Northwest Cloud Band shown schematically in Fig. 17.10(a) and in the satellite imagery in Fig. 17.10(b).

17.7 OTHER DEPRESSIONS

Many depressions in temperate regions, not just in the tropics, are without fronts or the fronts form later as a result of the convergent wind flow of the depression. This is shown in typical synoptic pattern in Fig. 17.11 where one sees the creation of a strong temperature gradient, and the reader will recognize that this is an example of the idealization in Fig. 17.8.

Other fronts form on the polar front, on troughs due to land/sea temperature contrasts (e.g. Western Australia, and off the east coasts of most continents in winter). The development of lows, with or without fronts, occurs in association with an upper trough.

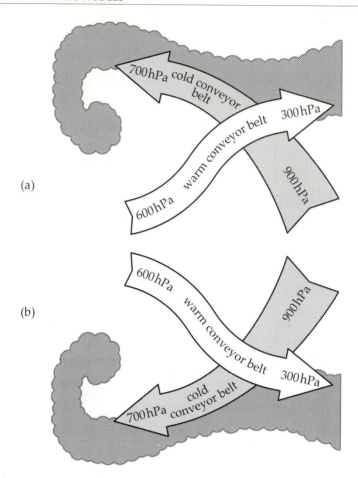

FIGURE 17.9 Browning model of cold and warm conveyor belts associated with an occlusion and cloud band: (a) northern hemisphere; (b) southern hemisphere.

Another important feature is the cut-off low. This is a low which becomes 'trapped' in the mean easterly flow, in either hemisphere, often with a substantial ridge extending poleward. In the right circumstances with a suitable stable upper flow configuration and topographic 'anchoring', these coupled systems, usually referred to as blocking pairs and sometimes as dipoles, persist for many days. Their most obvious effect is to prevent the commonly expected eastward movement of lows; such lows tend to weaken as they approach the block and slide poleward of the ridge. However, sometimes a low or an associated secondary, or an associated upper low, will be attracted into the cyclonic

FIGURE 17.10 (a) Schematic model of the three-dimensional flow in a northwest cloud band. L, M and H refer to the levels of isentropic streamlines (from Ferrière, 1994 and BMTC, 1988). (b) Satellite image and m.s.l. pressure pattern for a typical north west cloud band. (Image courtesy Bureau of Meteorology from Japanese GMS data.)

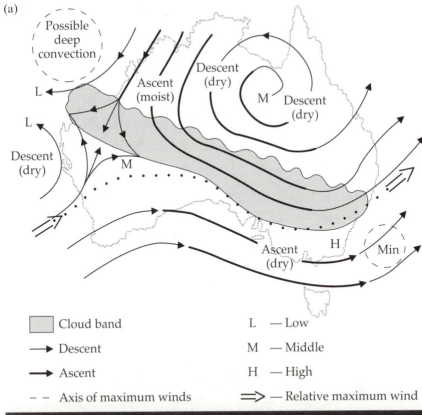

(a)

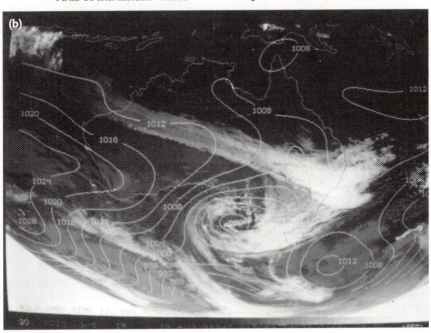

(b)

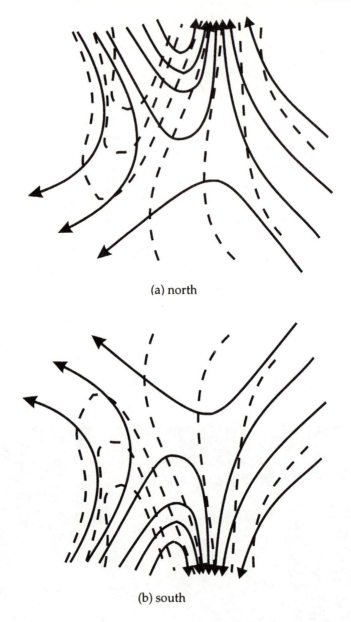

(a) north

(b) south

FIGURE 17.11 Typical synoptic surface wind isotherm patterns (full and dashed lines respectively) associated with intensification of a low: (a) northern; (b) southern hemisphere.

portion of the block – tracking equatorward and eastward. This process, known as 'shearing off' by forecasters, effectively reinforces the cyclonic portion and tends to prolong the existence of the blocking pair.

Depressions are associated with upslide and can result in widespread stratiform cloud. Depending on the hydrostatic stability, embedded cumuliform cloud may form. Thus rain and showers are possible even before a low has developed as a cyclonic circulation on the m.s.l. analysis.

Troughs, depressions and fronts may provide lifting of up to 300 hPa in a 24 hour period. Given a moisture supply, such lifted air is likely to form stratiform cloud and to precipitate in the form of rain or snow. Up motion may also be provided by mountains, and this too will tend to produce stratiform cloud. In contrast, up motion of a convective or cumuliform nature may be triggered by the upslide lift and/or by thermal convection provided that the atmosphere is conditionally or potentially unstable.

Shower activity is expected in the colder stream just west of a trough or front. Further west where the low-level convergence is less, or possibly divergence may be occurring, any thermal convection will be limited by the subsidence aloft. Thus vast expanses of strato-cumulus may form on the poleward side of anticyclones. Ahead of a trough or front, the poleward-moving air can be potentially unstable through deep layers and so deep convection is possible. This air is relatively warm and if it has been over the sea it may contain much more precipitable water than the cooler equatorward moving air. Therefore thunderstorms, heavy showers (and flash flooding) are more likely in these prefrontal or pretrough streams.

17.8 STEERING AND DEVELOPMENT

We have seen that in the right circumstances, depressions in association with a temperature gradient can be self-developing. It is commonly observed that the winds in the middle troposphere, say 500 hPa or the 1000–500 thermal wind, appear to act as a steering flow for features on the surface chart. An empirical rule of thumb is that the surface systems move parallel to the 500 hPa contours with half the 500 hPa wind speed. It is important to understand that though it *appears* that surface features travel downwind like rafts on a river, in reality the surface feature is being continuously redeveloped in the region of the strongest isallobaric components. These rise/fall patterns may be regarded as mainly due to temperature advection and vorticity advection. Mid-level and upper level patterns also influence the rate of intensification or weakening of surface cyclones and anticyclones.

We have already derived a useful relationship for the diagnosis, but not prognosis, of synoptic systems. Recall the omega equation (equation (15.19)). The first and second terms are the vertical motion terms: the third term is called the thickness, or temperature, advection; and the fourth the differential vorticity advection. In the absence of other effects, warm advection implies up motion, cold advection implies down motion. Similarly, rising air is implied by an increase with height of cyclonic vorticity advection; subsiding air is implied by an increase with height of anticyclonic vorticity advection. This allows us to re-express the omega equation in words (following the treatment of Holton

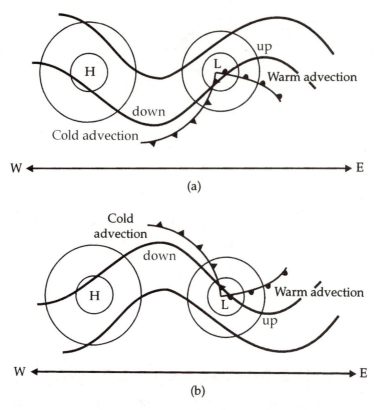

FIGURE 17.12 Vertical motion due to cold and warm advection as deduced from equation (17.7). M.s.l. pressure pattern represented by fine lines and 500 hPa pattern by thick lines: (a) northern; (b) southern hemisphere.

(1979)), thus:

Rising motion $\propto$ warm advection

+ vertical rate of increase of cyclonic vorticity advection

Sinking motion $\propto$ cold advection

+ vertical rate of increase of anticyclonic vorticity advection

$$(17.7)$$

This might appear complicated at first, but is quite easy and powerful once one works through a few examples. At all levels, vorticity advection in troughs, ridges and near the centres of synoptic features is very small since the gradient of vorticity is very small; conversely, vorticity advection, either cyclonic or anticyclonic, is usually greatest between such features. Thus Figs 17.12 and 17.13 show a schematic surface high and low beneath a 500 hPa flow. For the low, the cyclonic vorticity advection at the surface is roughly zero while directly aloft it will be comparatively large – hence, up motion is expected associated with falling

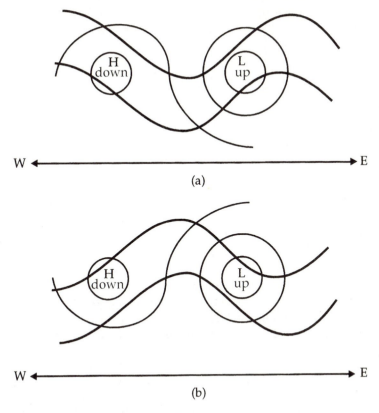

FIGURE 17.13 Vertical motion due to differential vorticity advection as deduced from equation (17.17).

surface pressures. Figure 17.13 shows the vertical motion due to differential vorticity advection.

Just after World War II, Sutcliffe (1947) developed a theory linking surface divergence with vorticity in the troposphere. His treatment, and that of several others, notably Petterssen (1956), gave forecasters the ability to undertake the diagnosis and prognosis of system development in qualitative and sometimes graphical ways. Sutcliffe used many of the assumptions underlying the derivation of the more modern omega equation (see Section 15.3) to arrive at the following:

$$-\nabla \cdot \mathbf{V}' = \frac{\mathbf{V}' \cdot \nabla}{f}(f + \zeta' + 2\zeta) \qquad (17.8)$$

or, in a form following the thermal wind,

$$-\nabla \cdot \mathbf{V}' = \frac{V'}{f}\left(\frac{\partial f}{\partial s} + \frac{\partial \zeta'}{\partial s} + 2\frac{\partial \zeta}{\partial s}\right) \qquad (17.9)$$

The prime denotes values in the 1000–500 hPa thickness layer; other values pertain to the 1000 hPa level but are assumed to approximate the surface values. The first term is the divergence in the thickness layer. Making the further assumption that 500 hPa is the level of non-divergence, then the surface wind divergence is given by

$$\nabla \cdot \mathbf{V} = \frac{V'}{f} \left(\frac{\partial f}{\partial s} + \frac{\partial \zeta'}{\partial s} + 2 \frac{\partial \zeta}{\partial s} \right) \tag{17.10}$$

Remember that *low-level convergence* (or negative divergence) implies rising air and falling pressure and falling heights. The third term in brackets, Sutcliffe called the *steering* term. Thus, the thermal wind V' advects surface vorticity. This is why the thermal wind (or the 500 hPa wind, which is a crude approximation) appears to steer the surface features. The second term is a *development* term due to the advection, by the thermal wind, of the 'thermal vorticity' which is proportional to the Laplacian of the thickness. The first term is the *beta* term or *latitude* term met in the Rossby wave equation and is usually comparatively small. Charney (1948) produced a rigorous scale analysis of all the terms used or implicitly discounted in the above.

It is suggested that the reader refer to Figs 17.12 and 17.13, and determine the contributions to low-level convergence by the advection term and by the development term (approximate the thickness by the 500 hPa flow). Compared with the omega equation, Sutcliffe's steering term contribution is similar to the differential vorticity advection term, and his development term is similar to the temperature advection term.

Several similar models and perspectives have been proposed by others. We saw in a previous section the advantage of a coordinate system fixed to the moving front. In such a system the vertical motion can be ascribed entirely to temperature advection. More recently, Hoskins and Pedder (1980) have developed a theory of Q vectors (first proposed by Sawyer and Eliassen) whereby the omega equation can be expressed by a single term – the convergence of the Q vector field.

Isotach analysis of the middle and upper level flow shows minima and maxima in wind speed. A jet max is usually regarded as any such maximum greater than 60 knots. Intuitively, the reader might expect jet max to be pertinent to surface development simply because of their energy. Jet max occur in various configurations but a very common form is ellipse shaped with the long axis along the wind. Jet max tend to move with the wind but at a slower speed. Forecasters often refer to a jet max as migrating around a trough axis.

To understand the effect of a jet max upon the underlying surface pattern, we consider the simple jet max configuration of Fig. 17.14. Maximum cyclonic vorticity is located just poleward of the jet max centre with maximum anticyclonic vorticity being just eastward. The gradient of vorticity in the direction of flow at these locations is near zero. Recognizing this, it is apparent that cyclonic vorticity advection occurs at jet stream level in the forward/poleward and rearward/equatorward quadrants. Thus in these quadrants – all other things being equal – there will be rising motion and associated low-level convergence possibly leading to surface low development. Conversely, in

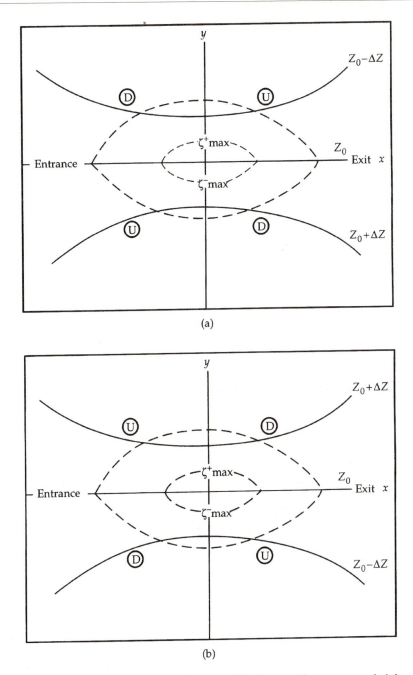

(a)

(b)

Figure 17.14 Model of an upper level jet max. Fine lines indicate contour heights of 250 hPa and dashed lines show isotachs (speed contours). Cyclonic and anticyclonic vorticity is a maximum at the indicated locations. Through equation (17.7), the vertical motion, U = up and D = down, is deduced. (a) Northern and (b) southern hemisphere.

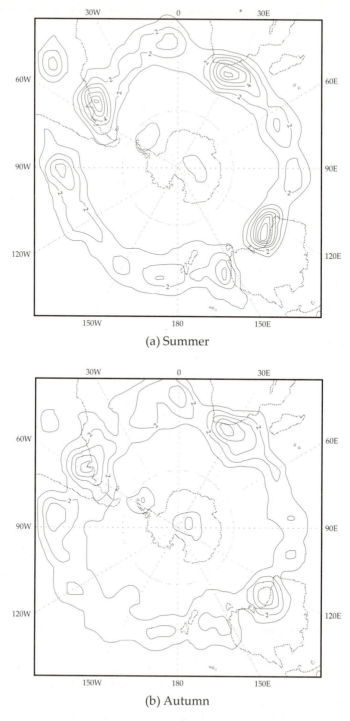

(a) Summer

(b) Autumn

FIGURE 17.15 The seasonally averaged density of anticyclogenesis. Contour intervals of 10^{-4} anticyclones per degree squared per day. (From Jones, 1994.)

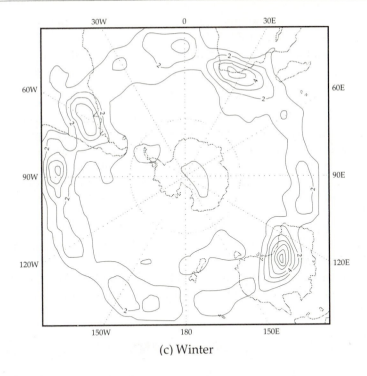

(c) Winter

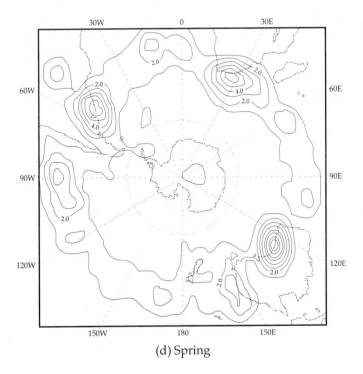

(d) Spring

the other quadrants, subsiding motion is expected with the formation of anticyclones being favoured.

The four-quadrant model was initially used by Riehl to show the associated development in the context of temperature advection. Of course, there are many other configurations possible but the ideas are transferable.

Vorticity associated with the jet max in Fig. 17.14 is due to shear and to curvature (recall Sections 11.7 and 11.8). By both Sutcliffe's method and the omega equation, the greatest cyclonic development of an underlying surface low tends to occur where the upper level cyclonic vorticity advection is greatest – all other things being equal, such as low-level vorticity advection and temperature advection. Thus, each quadrant may be labelled with U or D (for up or down motion within the column). Columns with up motion will favour cyclogenesis; down motion will favour anticyclogenesis.

An extreme form of cyclogenesis occurs in maritime 'bombs.' This is termed *explosive cyclogenesis* and defined as occurring when the pressure falls by more than 12 hPa in 12 hours. It occurs mostly in winter, a few hundred kilometres downstream of the mid-level trough (as in Figs 17.12 and 17.13), over the sea off the continental eastern coasts where there is often a strong cold-land–warm-sea contrast.

17.9 BLOCKING

Long-wavelength Rossby waves (also known as long waves or planetary waves) may be stationary or even slightly regressive as we saw in Chapter 12. Large-amplitude waves may form large-scale eddies which are evident throughout most of the troposphere. Consequently, the associated thickness patterns can be quite involuted with the steering direction for surface features being meridional. In the troughs, and especially in the ridges, of the long waves the thermal steering is weak and consequently the surface features move only slowly. With the large amplitude of the Rossby waves, the *beta* term in equation (17.8) is no longer negligible and tends to weaken the development of highs in an equatorward steering current, and the development of lows in a poleward steering current – that is, lows tend to weaken and 'slide' poleward along the western flank of the anticyclonic portion of the long wave.

We saw in the frontal theory section that mesoscale interaction can amplify the synoptic-scale features. A process by which the synoptic features can amplify the long waves also occurs although the mechanisms are less well understood. Typical lifetimes are 1 to 4 weeks in the northern hemisphere but significantly less, about a week, in the southern hemisphere. Particular locations and longitudes seem to be favoured. Seasonal densities of anticyclogenesis (the formation or intensification of highs) for the southern hemisphere, taken from Jones (1994), are shown in Fig. 17.15. These observations indicate that some aspect of topography is important. Large anticyclonic anomalies seem to occur more often than cyclonic ones. As mentioned in the Section 17.7, blocks of a dipolar nature also occur.

17.10 TROPICS

Although frontal activity is rare in the tropics, widespread weather in zones of convergence is not. Within the so-called equatorial trough, once known as the doldrums, local convection is the main type of weather, although in southeast Asia, smoke and haze is becoming an increasing problem. Convergence zones within the equatorial trough may lead to more widespread cloud and precipitation. Convergence occurs on an even larger scale if the trade winds from either hemisphere meet in a narrow zone, the intertropical convergence zone (ITCZ). This zone produces extremely bad weather conditions over an area several hundred kilometres in width.

Tropical disturbances can be found on daily charts. These include various types of eddies, vortices, troughs in the easterlies (typically a few thousand kilometres apart and travelling westward at 15 knots), other wave patterns (e.g. the 40 day oscillation), trade wind surges and convergence zones. The term 'disturbances' refers to any feature that disturbs the basic or mean flow patterns. The Ekman layer wind profile (see equation (14.14)) shows that wind in the friction layer actually blows cross-isobarically into the low pressure. This secondary circulation can be seen clearly in a simple analogue demonstration. To a glass flask of water add some tea leaves. Stir or swirl vigorously. Radial inflow takes place at the bottom. Occasional 'pumpings', or short-lived towers, of the leaves may also be evident.

This convergent inflow is unstable (in the sense of self-reinforcing) for an atmosphere which is neutrally buoyant or potentially unstable. This latter condition applies to much of the tropics. The important mechanism for producing up motion is known as CISK (Conditional Instability of the Second Kind). Even though the mechanism relies on the release of latent heat to fuel the development of a cyclonic disturbance, most of these disturbances are cool cored. That is to say, the mid-level temperatures of the disturbances are cooler than the environment – as are their mid-latitude counterparts. The tropical disturbances, however, may be punctuated by many hot towers which may be warm cored on a smaller scale. There is, though, one important tropical disturbance which is warm cored – the tropical cyclone.

17.11 PROBLEMS

1. Why might forecasters be especially concerned to monitor the migration of a jet max around an upper level trough located a few hundred kilometres west of the east coast?
2. From your local weather service obtain the routine charts (analyses) for a particular day of surface pressure, thickness, 500 hPa and 250 hPa or similar, and the 24 hour surface prognostic. Explain qualitatively the prognosticated movements of the various highs and lows and any development or decay.
3. If you believe any of the following lore from temperate latitudes, explain their possible basis in terms of synoptic features.

(a) Red at night – shepherd's delight
Red in the morning – shepherd take warning!
(b) A halo around the moon means steady rain the next day.
(c) A heavy morning dew suggests a fine and sunny day.
(d) If corns, rheumatics or old scars start nagging, a wet spell isn't far away.

THE TROPICAL CYCLONE

The most vigorous of the tropical disturbances are the tropical cyclones, also known as hurricanes, tropical storms or typhoons. As soon as a tropical disturbance has an observed, or reliably estimated, sustained surface wind (averaged over a period of 10 minutes) of $17\,\mathrm{m\,s}^{-1}$ (34 knots) or more it is defined as a tropical cyclone, although not all countries use this definition. Although regional usage of the terms is variable, World Meteorological Organization publications refer to tropical storms as being tropical cyclones with closed isobars and maximum sustained wind between 17 and $32\,\mathrm{m\,s}^{-1}$ (34 and 63 knots), and typhoons and hurricanes as tropical cyclones with maximum sustained winds of at least $33\,\mathrm{m\,s}^{-1}$ (64 knots).

The word cyclone comes from the Greek word *kukloma* meaning coiled snake and was proposed by Piddington in 1848. The word hurricane is thought to derive from a central American tribe's name for *god of evil*. The origin of the word typhoon is less certain. Possibly it derives from Chinese dialects for *great wind*. The practice of naming tropical cyclones seems to have begun at the turn of the century with Clement Wragge, a rather controversial and colourful government meteorologist in Australia. He used Greek letters, mythological characters and then the names of politicians, especially those he disliked. He published accompanying notes on the behaviour of the cyclones such as 'erratic', 'capricious' and 'very destructive' in a mischievous manner. Not surprisingly, he himself earned the nickname 'Inclement'.

A tropical cyclone first appears on the m.s.l. pressure chart as a tropical disturbance. Under favourable conditions its central pressure decreases rapidly, frequently to less than 960 hPa, with the lowest recorded being 870 hPa. The diameter of a tropical cyclone is typically a few hundred kilometres. Extremely steep pressure gradients result and the winds in quasi-gradient balance may reach hurricane force, with torrential rain and thunderstorm activity. Tropical cyclones usually move with a deep-layer-steering environmental wind. They may move in any direction and are sometimes quite erratic, although in the mean they

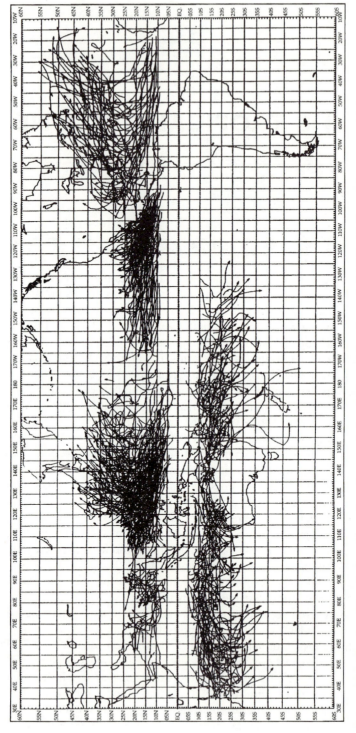

Figure 18.1 Tracks of tropical cyclones (maximum winds >34 knots) for the period 1979–88 (from Neumann, 1993)

move westward and poleward before recurving to move poleward and eastward. Figure 18.1 shows tropical cyclone tracks and genesis positions.

18.2 Structure and Energy Source

From a satellite perspective (Fig. 18.2), the striking features of a tropical cyclone are the swirl of dense cirrus cloud covering hundreds of kilometres, sometimes an eye, and the spiral arms of convective cloud. The typical scale and structure is shown in schematic form in Fig. 18.3 and the temperature and wind speed profile is similarly shown in Fig. 18.4. An important feature is that higher temperatures occur in the core, that is in the vicinity of the eye and its wall. Compared with other depressions, even gale-producing monsoon depressions, the tropical cyclone is *warm cored*.

These schematics assume an axisymmetric structure but this is not strictly true. Much of the asymmetry is in fact due to the superposition of the environmental flow upon the vortex. A practical outcome of this for mariners is

Figure 18.2 Satellite picture of tropical cyclone *Frank*, 11 December 1995. Courtesy Australian Bureau of Meteorology from Japan Met Service GMS satellite data.

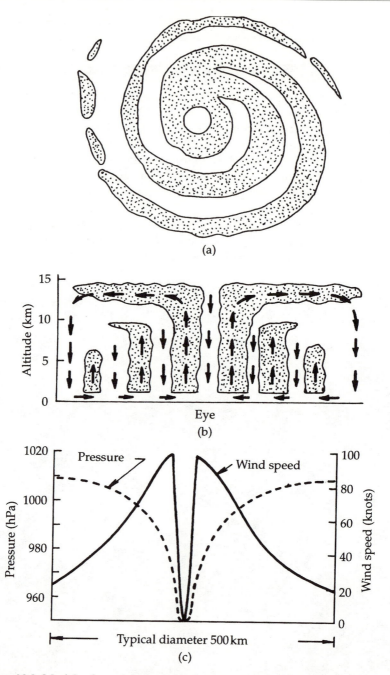

FIGURE 18.3 Model of a typical tropical cyclone: (a) plan view (in northern hemisphere) of spiral cloud bands – note the central eye; (b) vertical 'slice' showing the radial circulations which are superimposed upon the much greater tangential horizontal wind speeds; (c) tangential wind speeds at the surface with corresponding m.s.l. pressure for (b).

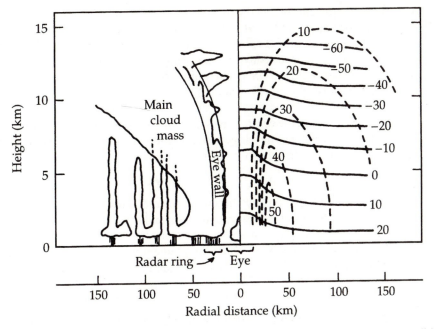

FIGURE 18.4 Vertical slice through a hurricane showing temperature structure (solid lines) in relation to tangential speeds (dashed lines are isotachs in m s^{-1}) and various cloud features (from McIlveen, 1992)

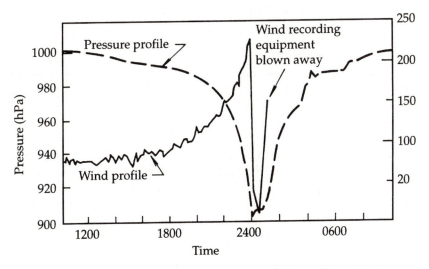

FIGURE 18.5 Pressure and wind recordings from an offshore oil rig with the passage of tropical cyclone *Orson*. Courtesy Australian Bureau of Meteorology from data supplied by Woodside Offshore Petroleum.

that in the northern hemisphere, the right leading quadrant following the cyclone track has the strongest winds (left in the southern hemisphere). An example of the wind and pressure profile for a tropical cyclone is shown in Fig. 18.5 for tropical cyclone *Orson* as measured at an offshore oil platform.

In countries other than the USA, where aircraft data are not usually available, the central pressure of a tropical cyclone is usually estimated from infrared and visual satellite imagery using a scheme devised by Dvorak (1975, 1984). This empirical scheme relates the central pressure to features of cloud configurations and size. Holland (1982) has developed a semi-empirical formula for the surface pressure p at a radial distance r from the centre of the eye. Holland's formula is

$$p = p_c + (p_n - p_c)\exp\left(-R/r\right)^b \qquad (18.1)$$

where p_n and p_c are respectively the environmental pressure away from the influence of the cyclone and the central pressure, and b is a factor between 1 and 2.5. At this stage, R is simply a suitable scaling constant, but we will soon see that it has physical significance. Now, using the gradient wind equation (8.12), the tangential wind speed is a function of $\partial p/\partial r$. Thus

$$v^2 = [b/\rho(R/r)^b(p_n - p_c)\exp\left(-R/r\right)^b - r^2 f^2/4]^{1/2} - rf/2 \qquad (18.2)$$

Using the cyclostrophic approximation (equation (8.10)) at $r = R$, the maximum speed V is given by

$$V^2 = b(p_n - p_c)/(\rho e) \qquad (18.3)$$

where ρ is the density and e is the base of natural logarithms. This implies a maximum wind speed at some distance from the centre, R (the radius of maximum wind) which is usually about 8 km within the eye wall. To estimate b, Love and Murphy (1987) suggest the following:

$$b \approx 0.25 + 0.30\ln(p_n - p_c) \qquad (18.4)$$

where $(p_n - p_c)$ is in hPa. Thus one needs to know only the central pressure and the environmental pressure in order to estimate the pressure and wind speed profiles. Love and Murphy suggest a further correction to the estimate of V such that corrected V (denoted V') is given by

$$V' = V + 0.0039V^2 \qquad (18.5)$$

This estimate is for the gradient wind which is above the friction layer, loosely assumed to be about 1000 m. Note that this is consistent with the height of the maximum tangential speed in Fig. 18.4. The 10 minute mean surface wind over the sea is often assumed to be 70% of the gradient wind speed. Other maximum wind speed formulae are

Atkinson/Holliday	$V = 3.4(1010 - p_c)^{0.644}$	(18.6)
Takehashi	$V = 6.9(1010 - p_c)^{0.5}$	(18.7)
Kraft	$V = 6.3(1010 - p_c)^{0.5}$	(18.8)
Fujita	$V = (13.1/a)(1010 - p_c)^{0.5} \quad 1 \leqslant a \leqslant 5$	(18.9)

For equations (18.6) to (18.9), wind speed is in m s^{-1} and pressure is in hPa. Note that equation (18.3) is, of course, in SI units. By setting p_n to 1010 hPa, the reader can check that all these equations are consistent.

Subsequent practical problems are forecasting the 'open-water' wave height created by tropical cyclones and the surge in sea level at landfall. The maximum significant wave height, H in m, due to a tropical cyclone in deep water is given by Hsu's empirical formula

$$H = 0.2(p_n - p_c) \tag{18.10}$$

where pressure is in hPa. Surge heights are given by similar rules of thumb for various locations. The current practice is to build up a library of scenarios using numerical models and to have these scenarios available operationally.

In previous chapters, we saw that the thermal wind equation implies that cyclonic wind speed increases with height around a cold-centred low-pressure region. Using the cyclostrophic equation (8.10), we can derive an analogous 'thermal' wind equation similar to that in Section 13.3.

Starting with equation (8.10), then

$$v^2 = \frac{r}{\rho} \frac{\partial p}{\partial r} \tag{18.11}$$

or, recasting into the isobaric system, we have

$$v^2 = rg \frac{\partial z}{\partial r} \tag{18.12}$$

Applying equation (18.12) to isobaric heights denoted by subscripts 1 and 2 and subtracting one from the other, then

$$v_2^2 - v_1^2 = rg \frac{\partial(z_2 - z_1)}{\partial r} \tag{18.13}$$

We can integrate the hydrostatic equation (15.8) to obtain

$$g(z_2 - z_1) = -RT \ln(p_2/p_1) \tag{18.14}$$

where T is the mean temperature between levels 1 and 2. So

$$\frac{v_2^2 - v_1^2}{z_2 - z_1} = rR \ln\left(\frac{p_1}{p_2}\right) \frac{\partial T}{\partial r} \tag{18.15}$$

$$= \frac{rg}{T} \frac{\partial T}{\partial r} \tag{18.16}$$

and if the layer is considered infinitesimal, then

$$\frac{\partial v^2}{\partial z} = \frac{rg}{T} \frac{\partial T}{\partial r} \tag{18.17}$$

In tropical cyclones the core is relatively warm and it is evident from Fig. 18.4 that $\partial T/\partial r$ is strongly negative in the mid-troposphere. The implication then of equation (18.17) is that the cyclonic wind speed is *decreasing* with height because this is a *warm*-cored system. The cyclonic wind speed decreases and reverses to form anticyclonic outflow aloft. Thus the system can have low-level inflow located beneath upper level outflow. The linking upward motion at mid-level in the eye wall has been recorded at $20 \, \text{m s}^{-1}$ or more.

The main source of this energy is the diabatic heating from latent heat of condensation. A second-order contribution is the diabatic heating from the sea surface. This occurs as the pressure of the inflowing low-level air reduces, thereby tending to cool the inflowing air adiabatically; however, any cooling is countered by sensible heat drawn from the warm sea surface. Most of the latent heat that drives the primary circulation is derived from evaporation from the sea surface located within a radius of 12° latitude degrees (Frank, 1987). Evaporation increases with reduced pressure and high winds. (Clearly a corollary of this is that tropical cyclones dissipate over dry land.) A model mechanism to feed in the energy is the CISK process which was briefly mentioned in the previous chapter. The process is shown schematically in Fig. 18.6 That process is now referred to as linear CISK in distinction to wave CISK. Under the linear CISK theory, the growth of cumulus clouds is predicted because the maximum growth rates occur at the smaller scales. Linear CISK can maintain and intensify existing tropical cyclones, but it cannot initiate them. For tropical cyclones to occur, there must be an effective interactive coupling between the cumulus-scale vertical motion and the large-scale horizontal wind field. The hypothesis of wave CISK is an attempt to match these scales of vertical and horizontal motion so that self-sustaining growth occurs. In the model of wave CISK, a wave, which could be a synoptic-scale wave in the easterlies, for example, with its associated upward motion is a precondition. With the wave, the zone of vertical motion slopes with height. The low-level vertical motion initiates moist convection and latent heat release. It has been shown that because of the sloping of the wave with height, the maximum heating is out of phase with the maximum upward motion and this leads to greater vertical motion at low levels and, thus, to self-amplification. Recently, Fraedrich and McBride (1995) proposed a third type of CISK based on the large-scale convective overturning in which the heating and the synoptic-scale vertical motion are related, or matched, to the cumulo-nimbus mass flux and the synoptic-scale mass flux. In their model, positive feedback occurs for motion of the order of several hundred kilometres, and does not occur for sea surface temperatures below 25.5°C, a threshold reasonably consistent with Gray's criteria (next section).

Section 5.6 introduced the concept of static stability (in relation to vertical displacements) in the form of the Brunt–Väisälä frequency, N. An analogous concept of the inertial stability in relation to horizontal perturbations is the inertial frequency, I (Holton, 1979). For horizontal flow in gradient balance (i.e. the pressure gradient, the centrifugal and Coriolis forces cancel), we simply assert here (but see Holland (1987) for a derivation) that

$$I^2 = (f + \zeta)(\zeta + 2v/r) \qquad (18.18)$$

and

- if $I^2 > 0$ then a stable oscillation with frequency I results;
- if $I^2 < 0$ then any displacement results in acceleration in the direction of the displacement;

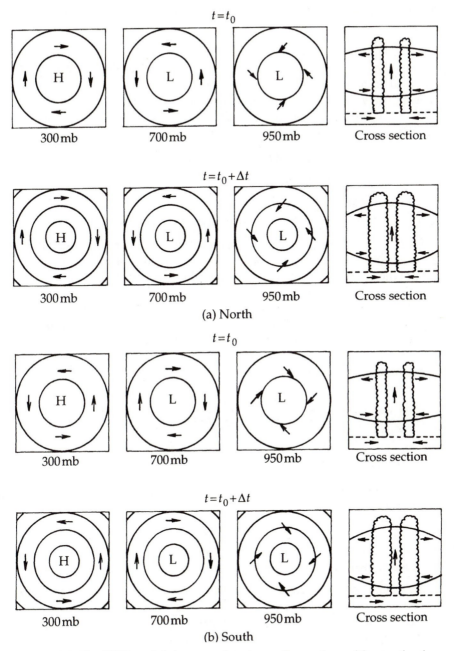

FIGURE 18.6 The CISK model. At $t = t_0$, there is a surface cyclone with an anticyclone aloft. Frictional convergence near the surface provides moisture and latent heat. Most of this diabatic heating is balanced by adiabatic cooling associated with induced radial circulation. However, at $t = t_0 + \Delta t$ there is a slight warming and accompanying thickness increase inducing a more intense upper anticyclone, a more intense surface cyclone and so on. (a) Northern and (b) southern hemisphere. From Frank (1987).

■ if $I^2 = 0$ then stability is neutral (see Schubert and Hack (1982) for a derivation).

Let H be the height scale, say 10 km, being the height of convective clouds, and let L be a horizontal distance scale for a cluster of such clouds. Now NH is the speed of the gravity waves which disperse the diabatic heating in the clouds: heuristically, LI may be thought of as the speed of inertial response of the wind field to rearrange the mass field (i.e. the pressure field). Let F_R be defined as LI/NH. F_R is called the rotational Froude number and

$$F_R = L(f + \zeta)^{1/2}(\zeta + 2v/r)^{1/2}/NH \qquad (18.19)$$

According to the value of the Froude number the response is such that

■ if $F_R \ll 1$ then the pressure field adjusts to the wind field and
■ if $F_R \gg 1$ then the wind field adjusts to the pressure field.

But if $F_R \sim 1$ then the pressure and wind fields interact with each other; that is, the scales are matched. For a tropical convective cloud cluster of 100 km extent, F_R is of the order of 0.01 and thus the cloud clusters are inefficient at converting their latent heat release to warming which would decrease the pressure and increase the rotational circulation. Instead, much of the energy is dispersed by gravity waves. Inspection of equation (18.19) shows that clusters can be more efficient in a region of higher I, in turn due to an unusually high value of relative vorticity, ζ. (Remember that ζ is composed of both a shear term and a curvature term, so we might expect pre-existing troughs and shear lines to be genesis factors.)

18.3 GENESIS

A climatological survey by Gray (1968) showed that tropical cyclone genesis is related to the following six factors:

1. above-average low-level vorticity;
2. middle-level moisture;
3. conditional instability through a deep layer;
4. a warm sea surface (>26.5°C) and a deep oceanic mixed layer;
5. weak vertical shear of the horizontal wind;
6. a location at least a few degrees from the equator.

Maps of Gray's factors coincide with maps of tropical cyclones (genesis and tracks) as shown in Fig. 18.1.

The dependence on sea surface temperature is shown by the empirical relationship in Fig. 18.7. Various surface and upper level features have been identified as tropical cyclone precursors. For example, genesis is favoured in the wake of a tropical cyclone but inhibited in the region ahead (Frank, 1987). Broad areas of upper level divergence are favourable to tropical cyclone genesis and such areas can be associated with tropical upper tropospheric troughs (TUTTs) and with entrances to upper level easterly jets. Westerly wind shear (in the vertical) inhibits genesis. Idealized surface charts (by Love (1985) in

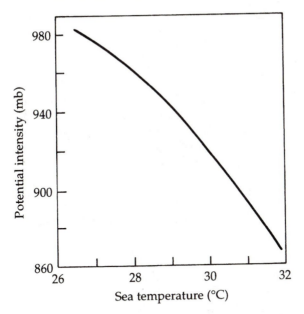

FIGURE 18.7 Empirical relationship of potential intensity of tropical cyclones to sea surface temperature of the genesis area.

Fig. 18.8) identify precursor synoptic features such as enhanced winds into the ITCZ (Intertropical Convergence Zone), a surge from a subtropical high, and the intrusion of mid-latitude fronts to low latitudes. In qualitative terms, such features serve to increase the ζ and v/r terms in equation (18.19) and thereby increase F_R, so that scale interaction is more likely.

An attractive conceptual model of tropical cyclone genesis proposed by Emanuel (1988) is that tropical cyclones are self-amplifying systems. They intensify to attain their MPI (Maximum Potential Intensity) *unless* their surroundings disrupt the process, as is usually the case. Strong vertical shear is the most common inhibitor. MPI is a function of sea surface temperature and the temperature of the outflow which is assumed to be at the tropopause, as shown in Fig. 18.9.

Gray, in particular, has pointed out that aggregate tropical cyclone genesis appears to fluctuate on several quasi-periodic scales owing to:

1. ENSO (El Nino Southern Oscillation) – the response of each basin is individual; for example, during El Nino, the North Atlantic basin has a large decrease in cyclone frequency whereas the South and Central Pacific basin has an increase.
2. QBO (Quasi-Biennial Oscillation) – this is a 26 monthly (roughly) oscillation in the equatorial stratospheric winds from an easterly phase to a westerly phase (*see* Fig. 18.10). As with the ENSO, the effect is basin dependent.
3. MJO (Madden–Julien Oscillation) – alternations of active and inactive genesis periods of 15 to 25 days have been observed with a 4:1 ratio of

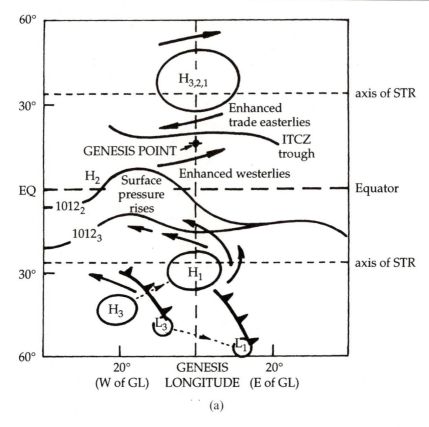

FIGURE 18.8 Schematic surface chart showing the important synoptic features 3 days and 1 day before cyclone genesis in (a) the northern and (b) the southern hemispheres. Subscripts denote days before genesis. (From Love, 1985.)

cyclone numbers between active and inactive periods (*see* Fig. 18.11). The length of these active and inactive periods suggests a link with the 30 to 50 day cycle, usually termed the Madden–Julien Oscillation.

18.4 STEERING AND DEVELOPMENT

Tropical cyclone positions are determined by a range of methods: satellite image interpretation, by radar, by aircraft and by analysis of available synoptic observations. Even so, accurate position fixing of tropical cyclones is difficult. Examples of their erratic, often trochoidal, motion at the smaller scales are shown for the paths of *Joy* and *Rewa* in Fig. 18.12. In seeking to understand the motion and development of tropical cyclones, forecasters use a variety of concepts. These are the beta effect, steering, empiricism, the Fujiwhara effect, a range of numerical models and statistical techniques.

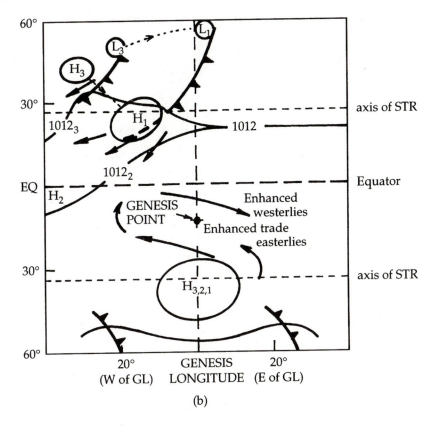

(b)

FIGURE 18.8 Continued.

18.4.1 Movement

Beta effect

In the previous chapter we inferred (mid-latitude) cyclone movement from the diagnosis of the vorticity tendency. We now make the basic assumption that a tropical cyclone will be displaced towards the region with greatest maximum cyclonic vorticity tendency. For an initially stationary axisymmetric tropical cyclone experiencing no background flow, we show heuristically that the cyclone itself generates a westward and poleward movement. Using the vorticity–divergence relation of equation (11.10), we have

$$\partial\zeta/\partial t = -\mathbf{V}\cdot\nabla\zeta - \beta v - f\nabla\cdot\mathbf{V} - \zeta\nabla\cdot\mathbf{V} \qquad (18.20)$$

where $\mathbf{V}$ is the horizontal wind vector and v is its meridional component. We now consider the contribution of the individual right hand terms at some distance from the centre. A little thought reveals that we are only interested in the *asymmetric* or net contributions. Since at any given radius both ζ and the tangential components of $\mathbf{V}$ are constant, the first and fourth terms contribute symmetrically: this is easily realized if one uses an r, θ coordinate system. To the

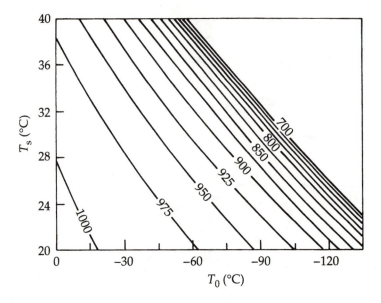

FIGURE 18.9 Maximum potential intensity as a function of sea surface temperature and outflow temperature at tropopause level. An ambient pressure of 1013 hPa and relative humidity of 80% are assumed. (From Emanuel, 1988.)

west of the cyclone, the $-\beta v$ term contributes cyclonic vorticity; to the east, it contributes anticyclonic vorticity. Thus we expect westward motion in either hemisphere. This is called the linear beta effect. If f is constant, the third term's contribution would be symmetrical. For the convergence associated with the cyclone, $-\nabla \cdot \mathbf{V}$ is positive. Poleward of the cyclone centre f is numerically greater, albeit slightly, and equatorward it is numerically smaller. Thus, on the

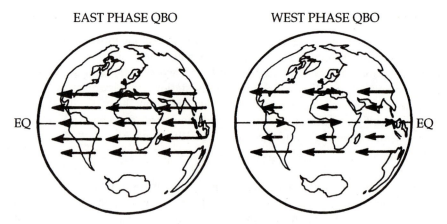

FIGURE 18.10 Schematic of the two phases of the QBO, having either easterly or westerly winds above the equator at 50 hPa. Depending on the basin, the particular phase suppresses or enhances tropical cyclone activity.

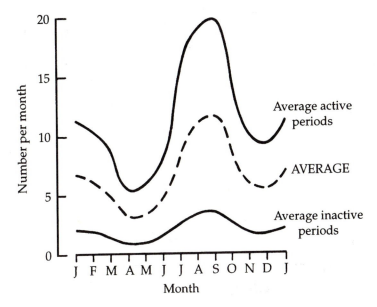

Figure 18.11 Effect of the MJO on tropical cyclone genesis. Average annual variation shown by dashed line with average variations for active (upper) and inactive (lower line). (From WMO, 1995.)

poleward side there is an asymmetric contribution of cyclonic vorticity (and anticyclonic vorticity on the equatorward side). In total then, the movement expected is to the west and poleward. The above exposition has the advantage of simplicity, although the reader should be aware that more rigorous arguments incorporating non-linear beta effects are usually called upon to explain the poleward motion.

Steering

Just as the mid-level wind or the thermal wind apparently steers mid-latitude systems, it has been observed that tropical cyclones tend to move with the speed and direction of the mid-level environmental wind. Some offices use the 700 or 500 hPa wind. Holland recommends a layer mean between 850 and 500 hPa filtered to remove the cyclone scales, but cautions that the notion is too simplistic as interaction between the tropical cyclone and the environment has a marked impact upon the track of the tropical cyclone. In general, the steering flow is easterly near the equator and becomes westerly with latitude. Bearing in mind the beta effects and the climatological broad-scale environmental flow, it is not surprising that the mean tropical cyclone tracks are initially westward prior to an anticyclonic recurving away from the equator.

Empirical rules

Recognizing synoptic patterns that favour certain types of motion and development is the next method. For example, Foley and Hanstrum (1994) developed a

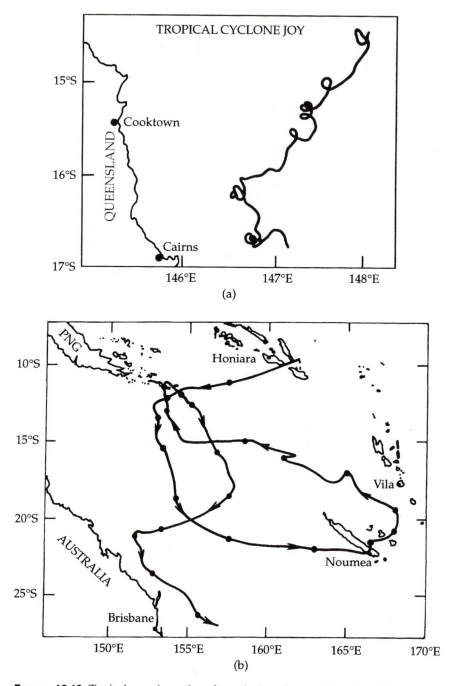

FIGURE 18.12 Typical erratic paths of tropical cyclones: (a) path of *Joy* 21–25 December 1990 heading towards the Australia coast; (b) path of *Rewa* 29 December 1993 to 21 January 1994 from Honiara. (Courtesy Australian Bureau of Meteorology.)

synoptic classification to distinguish those patterns that favour a tropical cyclone moving steadily down the West Australian coast and those that accelerate. A commonly observed pattern is that of the subtropical ridge poleward of the tropical cyclone: in this case the tropical cyclone will maintain its westward course. A trough in the upper level westerly flow to the west of a tropical cyclone usually indicates recurvature. Both of these empirical rules may be seen as consistent with the previous steering rules. Tropical cyclones tend to move towards the downstream end of convective cloud bands in the outer circle and give the impression of avoiding cumulo-nimbus-free sectors. A mass of middle-level cloud streaming poleward from a tropical cyclone indicates that recurvature is likely. 'Nearby' tropical cyclones mutually affect each other in complex ways and the interactions of binary tropical cyclones and of tropical cyclones with other systems is under vigorous research. It is often observed that tropical cyclones within 1000 to 1500 km of each other tend to inhibit each other.

Animation of satpic enables forecasters to identify and monitor the peripheral synoptic features that influence tropical cyclone motion and allow subjective forecasts. Dvorak (1975) proposed a set of pattern recognition techniques using such imagery for tropical cyclone motion but these are not widely used (Holland, 1993). (His techniques for tropical cyclone position-fixing and development are, however, widely used.)

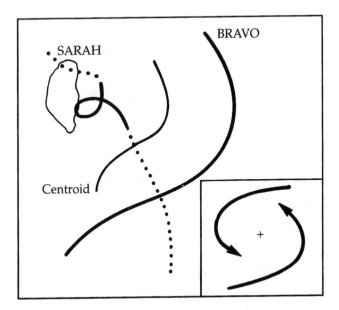

FIGURE 18.13 Component motion of a pair of binary tropical cyclones. Paths of cyclones *Bravo* and *Sarah* shown by heavy lines (dotted line indicating less reliable positioning). The motion can be resolved into mutual rotation about a common centroid as per inset and the motion of that centroid (shown by the fine line).

Fujiwhara effect

Tropical cyclones within 1500 km tend to rotate about each other, or more precisely, about a common 'centroid'. This effect is the Fujiwhara effect. What appears at first to be chaotic motion of two tropical cyclones may often be resolved into components of mutual rotation and of advection of the centroid on the broad scale. Figure 18.13 shows schematically the technique used to resolve the separate components. Tropical cyclone positions are plotted on an overlay map in a coordinate sysytem that moves relative to the centroid. The centroid is the geometric centre of the line joining the centres, sometimes weighted towards the larger cyclone. The motion of the individual cyclones may then be forecast by assuming persistence of the binary orbit and of the motion of the centroid.

Statistical methods

Simple track prediction schemes rely on climatology and persistence. Predictions based on climatology use the long-term mean track and speed of previous cyclones in the area, and those based on persistence use the current track and speed. However, subjective extrapolation using persistence is usually improved by smoothing the recent track using regression methods such as suitable order polynomials. Another form of climatological forecasting is an analogue technique where a 'family' of suitably selected storms is found from the archives and the forecaster is then guided by their known history. The optimal combination of the climatological and the persitence forecasts is referred to as CLIPER (CLImatology and PERsistence).

CLIPER provides simple and realistic forecasts of cyclone track. It also provides a reference by which to judge the accuracy of other (more sophisticated) techniques. These more sophisticated methods include dynamical methods which range from fairly simple trajectory forecasts using environmental advection implied from the large-scale flow predicted in turn from one of the simpler global numerical models, to the complex global numerical models integrated over several days. One problem with the latter is that they often incorrectly reposition the tropical cyclone during their initialization and assimilation phase. The position error at 'time = 0' is typically one or two hundred kilometres.

18.4.2 Development

Empirical rules

Tropical cyclones are intensified by the same factors that promote genesis. Examples include surges in the trade winds (easterlies) or the monsoonal westerlies, upper level features such as the development of an outflow jet, or a tropical upper tropospheric trough (TUTT) in a favourable position. These types of interactions form the basis for many of the empirical rules for forecasting intensification and are beyond the scope of this book.

A form of expert system widely used to estimate tropical cyclone intensity (and central pressure) and any tendency to intensify or weaken is the Dvorak scheme. Using satellite imagery, both visual and infrared, a T number is assigned to a tropical cyclone based upon the size, shape and evolution of features such as the eye, the spiral arms, the embedded cumulo-nimbus activity. This system works reasonably well and estimates of central pressure are usually within a few hectopascals.

Upper level divergence in the form of outflow jets at upper levels is extremely important to tropical cyclone development. Such jets are usually associated with the anticyclonic flow of the subtropical jet, with a TUTT, or with a trough in the westerlies. An adjacent cold pool may also provide an outflow region via the baroclinic jet around the cold pool.

Dissipation and interactions

Upon landfall, tropical cyclones weaken rapidly, mainly because of the loss of the latent heat sourced from the sea surface. However, heavy rain may persist for days. Cyclones can redevelop if the decayed system moves back over water. Tropical cyclones that move poleward of 30° latitude generally undergo extra-tropical dissipation and weaken as they move over land or cooler water, or encounter an environment of increasing westerly wind shear. Dissipating tropical cyclones will tend to maintain gale force winds within 2 kilometres of the surface for several days and these gales force winds can be brought down to the surface by vertical mixing due to convection or passage over rough terrain. However, about a quarter of tropical cyclones moving into mid-latitudes interact with a mid-latitude front and transform into an extra-tropical cyclone. During transformation, the cyclone may exhibit a mixture of tropical and extra-tropical characteristics. Compound extra-tropical transformation occurs when a tropical cyclone merges with a pre-existing extra-tropical cyclone which then intensifies owing to the addition of moisture and diabatic heating. Complex extra-tropical transformation occurs when a tropical cyclone approaches a front and induces a wave low on the front. The tropical cyclone then accelerates into the wave low and forms a single extra-tropical cyclone.

The changes from tropical cyclone to extra-tropical cyclone are summarized in Table 18.1.

18.5 FORECASTING SKILL

The performance of CLIPER forecasts provides a reference by which to judge the operational skill of other techniques. CLIPER forecasts also form the basis of Neumann's measure of forecast difficulty level (Neumann, 1993). The CLIPER errors provide a threshold skill level and are found to be directly related to operational forecast errors. Forecast difficulty level for the different basins varies a great deal. Figure 18.14 shows CLIPER forecast position error for the various basins at different lead times. For example, a 24 hour CLIPER forecast for a tropical cyclone in the southwest Indian Ocean has a mean error of 150 km. By comparison, an evaluation of the UKMO (United Kingdom Meteorological Office) global numerical model for 163 tropical cyclones over three seasons in the

Table 18.1 Structural and dynamical changes associated with extra-tropical transformation (from Merrill 1993)

Element	Tropical cyclone	Extra-tropical cyclone
Temperature	Axisymmetric, warm cored. Gradients strongest aloft.	Strongly asymmetric with fronts at low levels.
Wind	Mostly asymmetric and concentrated. Speed maxima typically to the right of the motion (left in SH).	Strongly asymmetric and less concentrated. Speed maxima in both cold and warm air masses, especially near fronts.
Rain	Somewhat symmetric, often heaviest to right of the motion (left in SH).	Asymmetric, often heaviest to poleward.
Energetics	Baroclinicity produced locally by evaporation from the sea in high wind and low pressure. Kinetic energy produced by concentrated, largely symmetric ascent in convective clouds in the warm core, subsidence in the eye and weak subsidence outside.	Concentration of existing large-scale baroclinicity by cyclone-scale advection. Kinetic energy produced by cyclone-scale ascent in the warm air and subsidence in the cold air. Condensation in the ascent region provides additional energy.

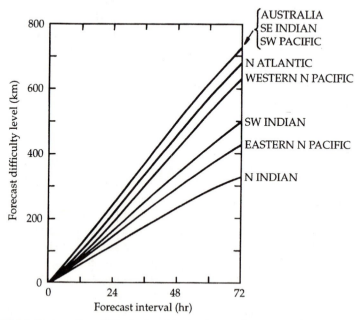

FIGURE 18.14 Forecasting difficulty level by basin as a function of forecast lead time (adapted from Neumann, 1993).

southwest Indian Ocean showed the model to have a mean error in forecasting position at 24 hours of 264 km. Much of this error (about 200 km) is due to a mislocation error at time $= 0$ during initialization. An interesting finding by F. Woodcock (personal communication) is that, at least for the Australian region,

$$FE = IE + 6.3\Delta t \qquad (18.21)$$

where *FE* and *IE* are the operational forecast and initial errors in kilometres and Δt is in hours. This finding implies that the error growth with lead time is independent of the initial error.

Forecasts of cyclones have certainly improved in recent years although much of the improvement is probably due to better position fixing and monitoring, in turn due to better observing technologies and communications.

18.6 PROBLEMS

1. From the pressure trace in Fig. 18.5, calculate an estimate of the maximum wind speed and compare it with the observed wind speed. At what distance from the centre of the eye is this speed attained?
2. (a) Assume SST (Sea Surface Temperature) of 28°C, T_o (outflow temperature) of -70°C. What is MPI? Using an environmental pressure of 1013 hPa, what is the maximum wind speed?
   (b) Plot specific humidity as a function of dew point. [Hint: Use a suitable formula from Section 4.21]. What is the relevance of this curve to Gray's fourth criterion for tropical cyclone genesis?
3. A day or so prior to the arrival of a tropical cyclone, the weather is often very fine with hardly a cloud in the sky. Explain.
4. A rule of thumb for forecasting recurvature is the presence of a mid-level westerly trough in mid-latitudes to the west of a tropical cyclone. Explain.
5. Examine Love's diagram (Fig. 18.7) and explain the effect of each feature in terms of (a) Gray's genesis factors and (b) the rotational Froude number.
6. Referring to Fig. 18.1, why are there no tropical cyclones observed over the ocean west of Peru?
7. A forecaster examines a series of satellite images of a tropical cyclone from an area without synoptic observations and makes an estimate of its maximum surface wind speed. Outline the forecaster's likely line of reasoning and calculations.

19

RADIANT ENERGY TRANSFER

19.1 HISTORICAL CONCEPTS, CAVITIES AND BLACK BODIES

The very existence of the earth and the maintenance of its important and life-sustaining atmospheric processes depends on radiant energy transfer. While the sun is the fundamental source of radiant energy, all exposed terrestrial surfaces and transparent or translucent fluids, particularly gases, participate in an interdependent and complex range of radiant processes.

The science of thermal radiation had its origins in the caloric theory in which hot and cold radiations were accepted concepts. Since a block of ice was thought to radiate cold to any body placed beside it, it is evident that these concepts were based on very subjective and shallow impressions. A simple further philosophical experiment, in which are considered the processes at play when water ice at 0°C faces dry ice (CO_2) at -80°C, highlights the arising dilemma as to which body is radiating what type of radiation.

Because of such logical problems, it was essential to return to the fundamental observation that even in a vacuum, thereby ruling out the possibility of heat transfer by conductive or convective means, two isolated bodies at initially different temperatures ultimately come to thermal equilibrium at the same temperature. From this basis it is possible to make progress with physical explanations.

Prevost put forward his theory of exchanges in 1792 to offer a logically satisfactory explanation of radiation transfer between surfaces at all temperatures. This theory, which might more accurately be called the theory of continual exchange, states that a body emits radiant energy at all temperatures and at a rate which increases with the temperature. The settling down of a body to a constant temperature then indicates the fact that it is receiving radiation from the surroundings at the same rate as it is emitting. Thermal equilibrium is thus dynamic rather than static and leads to the question of the relationship between the processes of emission and absorption.

A simple experiment was conducted by Leslie in 1804, using a hollow cube prepared with both reflecting and black-painted sides. When such a cube is filled with hot water, a qualitative test with the open palms of bare hands as sensors suffices to show that the black faces radiate more efficiently than the reflective ones. However, Ritchie in 1833 extended this experimental set-up definitively to allow for quantitative deductions which established both a fundamental fact as well as the precise definition of a black body.

Ritchie's experiment utilized Leslie's cube, again filled with hot water, but with two additional surfaces, one black and the other reflecting, able to be positioned parallel and equidistant from the faces of the cube. When a reflecting face is placed near to a black face on the cube and vice versa, it is observed, e.g. by means of a simple thermocouple or other thermometric arrangement as illustrated in Fig. 19.1, that no temperature difference develops between the two movable faces which have different surface finishes and are physically unconnected to the cube. This clearly indicates a link between the ability of a given surface to absorb and radiate respectively.

Historically, this led to the definition of a quantity known as the emissive power E, for a surface, as the energy per unit area per unit time emitted radiantly, and also the absorption coefficient or absorptivity, α, as the fraction of the incident radiant energy which is absorbed. While these parameters and their symbols are no longer in vogue, their use is essential in following the historical development of this subject. If in Ritchie's experiment, the black and reflecting surfaces are designated by the subscripts b and r respectively, then

$$E_r \alpha_b = E_b \alpha_r$$

$$\frac{E_r}{\alpha_r} = \frac{E_b}{\alpha_b} \tag{19.1}$$

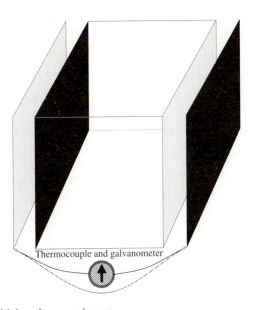

FIGURE 19.1 Ritchie's cube experiment.

Essentially, equation (19.1) implies that any surface is as good a radiator as it is an absorber, a statement which offers the most basic expression of the law first formulated by Kirchhoff:

A surface for which the absorptivity $\alpha = 1$, i.e. one which absorbs all incident radiation, is called a perfectly black body or, more simply, just a black body. For all other surfaces, $\alpha \leqslant 1$.

A further quantity, known as the emissivity ϵ of a surface, is defined as the ratio of its emissive power to that of a black body at the same temperature. It is clear that for a black body, $\epsilon = 1$, while for all others, $\epsilon \leqslant 1$. Then

$$\frac{\epsilon_r}{\alpha_r} E_b = \frac{\epsilon_b}{\alpha_b} E_b \tag{19.2}$$

Therefore, $\epsilon_r = \alpha_r$ and in general $\epsilon = \alpha$, since $\epsilon_b = \alpha_b$.

Equation (19.2) clearly demonstrates the equality of the two parameters, absorptivity and emissivity. The latter is now the scientifically preferred quantity.

In radiation theory, it soon becomes evident that many processes can readily be envisaged and that for many purposes the need for a technically complex experiment can be obviated by means of a controlled imagination. Exploring the relationship between the surface temperature of a radiator and the density of radiant energy offers an example.

Consider, as illustrated in Fig. 19.2, two enclosures, A and B, which are at the same temperature but whose surfaces, in being arbitrary, may be different. Assume that these two enclosures are connected by a tube equipped with a shutter able to block the transfer of radiation. If A were initially to have a greater radiation density than B, there would be a net transfer of radiation to B if the shutter were to be opened. Were the shutter to be closed again, new equilibrium conditions would ensue within the two separate enclosures. A, in having lost radiation, would cool down by giving radiant energy to the cavity until a new

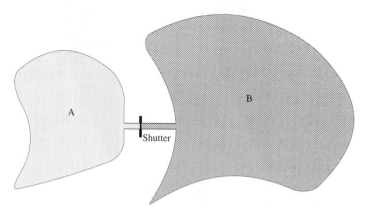

FIGURE 19.2 Two arbitrarily shaped, separate uniform temperature cavities connected by a tube and shutter.

equilibrium were reached. Correspondingly, the walls of B would rise in temperature to reach equilibrium with the increased radiation density. Clearly, through the generation of a temperature difference capable of powering an engine, even though the source and sink had initially been at the same temperature, the entire process outlined above would have violated the second law of thermodynamics.

It has thus been demonstrated that the energy density within an enclosure depends only on the temperature, and is independent of both the shape and the nature of the walls, as well as of the shape and nature of any bodies placed within it. Any body placed in an enclosure must take up the temperature of the enclosure and then remain in equilibrium. Under such conditions, the theory of exchanges explains that the enclosed body emits just as much energy as it receives from the enclosure walls, thereby leaving the radiation density within the enclosure unaffected.

The information above also allows a simple black body radiator to be designed, based on the concept of a thermodynamic cavity. If such a cavity is prepared with a hole through which radiation can enter or leave, it is clear that the plane of this aperture will have the properties of a perfectly black plane radiator, provided that the hole is small compared with the surface area of the cavity. This prescription means that only a minute fraction of radiation which is trapped after entrance through a cavity hole is able to escape; in other words, absorption is well nigh total.

While it is important and reassuring to know that in theory, at least, a perfectly black body radiator can be constructed, it is essential to devise criteria for the description of the radiative behaviour of any radiating plane surface. The cosine law, which is derived below, is one of the most fundamental.

Let K, the specific emission, be the rate of radiation per unit solid angle per

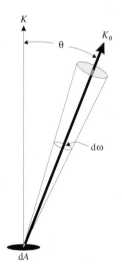

FIGURE 19.3 Radiation into an infinitesimal solid angle at various angles.

unit area of a surface in the normal direction and let the value of this quantity at an angle θ to the normal be K_θ. Consider the radiating surface element dA which is shown in Fig. 19.3, from which it can be seen that if dE_θ describes that fraction of the emissive power which radiates into an infinitesimal cone which subtends a solid angle $d\omega$ at the surface of the element dA and has its axis lying at an angle θ to the surface element's normal direction then

$$dE_\theta \, dA = K_\theta \, d\omega \, dA \qquad (19.3)$$

A second radiating surface element is now introduced as indicated in Fig. 19.4 and, to ensure radiative equilibrium, both are considered to be within the common confines of a uniform temperature enclosure. In order that this radiative equilibrium may be maintained, the radiation received by the element dA from dA' must equal that received by dA' from dA. Thus

$$K \, dA' \, d\omega' = K_\theta \, dA \, d\omega \qquad (19.4)$$

But since $dA' = r^2 \, d\omega$ and $dA \cos \theta = r^2 \, d\omega'$ it follows that

$$K \, dA' \, \frac{dA \cos \theta}{r^2} = K_\theta \, dA \, \frac{dA'}{r^2}$$

and therefore

$$K_\theta = K \cos \theta \qquad (19.5)$$

This is the basic expression of the cosine law, which was first formulated by Lambert for perfect optical diffusers (for which flat, fresh, light-scattering powder snow surfaces constitute a good example in nature) but which applies equally well to perfect radiators, which may be conveniently visualized as matt black surfaces.

The conveniently, if unusually, named quantity, the specific emission, which

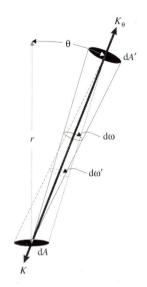

FIGURE 19.4 Radiative equilibrium between two infinitesimal surfaces.

was required for the demonstration of the cosine law, can be related to the emissive power in a manner which will be examined now.

Reference to Fig. 19.5 shows that the radiation, intercepted by a hemisphere, from an infinitesimal radiating element of area dA can be determined. If the emissive power of the element dA is E, and dE_θ describes that fraction of the emissive power which radiates into an infinitesimal cone which subtends a solid angle $d\omega$ at the surface of the element dA and has its axis lying at an angle θ to the surface element's normal direction, then, again,

$$dE_\theta \, dA = K_\theta \, dA \, d\omega$$

Lambert's law allows this to be written as

$$dE_\theta = 2\pi K \cos\theta \, d\theta$$

When this expression is integrated over the entire hemisphere above the radiating element, i.e. from $\theta = 0$ to $\pi/2$ as shown,

$$E = 2\pi K \int_0^{\pi/2} \sin\theta \, d(\sin\theta)$$

which reduces to

$$E = \pi K \tag{19.6}$$

This is an expression of fundamental importance.

As well as offering a comprehensible model for a radiative black body, a constant temperature enclosure or cavity provides a defined environment for examining other properties of black body radiation such as the density of radiant energy.

The energy density of radiation must be independent of position in a constant temperature enclosure or cavity, because this quantity has earlier been shown to be dependent only on the temperature. Figure 19.6 depicts a spherical cavity with its inner walls being that of a black body. At the centre of this cavity of radius r is shown an infinitesimal cylinder of length dl and cross-section $r^2 \, d\omega$ and hence volume $r^2 \, d\omega \, dl$.

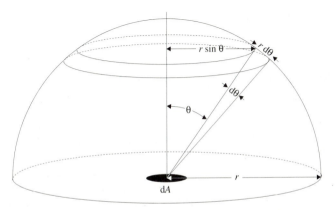

Figure 19.5 Radiation into a hemisphere from an infinitesimal black body.

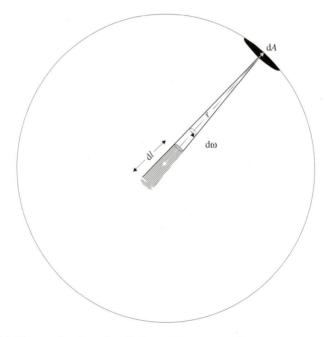

FIGURE 19.6 Energy density of radiation at the centre of a spherical cavity.

Assuming that the radiation emitted by the cavity wall element of area dA moves with a velocity c, it follows that at any instant, the quantity of energy at the centre of the sphere and originating from dA is $(K\,dA\,d\omega\,dl)/c$, where only normally emitted radiation need be considered, since radiation emitted in other directions bypasses the centre. It follows that the energy density of radiation at the sphere's centre, contributed by the wall element dA, is given by

$$du = \frac{K\,dA\,d\omega\,dl}{c\,dl\,r^2\,d\omega} = \frac{K\,dA}{cr^2} \tag{19.7}$$

If radiation from all directions, i.e. the entire inner surface of the spherical cavity, is considered, simple integration yields

$$u = \int_0^{4\pi} \frac{K\,dA}{cr^2} = \frac{4\pi Kr^2}{cr^2} = \frac{4\pi K}{c} \tag{19.8}$$

but on considering equation (19.6), it then also follows that

$$u = 4E/c \tag{19.9}$$

Equations (19.8) and (19.9) thus relate the four principal radiative parameters introduced so far, i.e. u, E, K and c, but one further basic parameter remains and that is the pressure exerted by radiation

Already in about the year 1600, the astronomer Kepler had suggested that solar radiation exerts a pressure which causes the tails of comets to appear to be blown away from the sun. In 1873, Maxwell showed that normally incident electromagnetic radiation exerts a pressure, on absorbing surfaces, equal to the energy density. No quantitative tests were made, however, until Lebedev in 1900 and Nichols and Hull in 1903 carried out laboratory experiments to verify this effect.

A simple analysis of the radiative pressure effect relies on analogy with the molecular case explained by the kinetic theory of gases. In this latter theory, the pressure exerted by the molecules of a gas against the walls of a container is proportional to the rate of change of momentum density at the wall.

The change of momentum per unit time per unit area of wall by molecules with collective density n molecules/(unit volume) and individual mass m is given by $\frac{1}{6}(2mv \cdot nv)$ because when resolved into three orthogonal directions, each with the two possibilities of either positive or negative motion, only one-sixth of the total can be regarded as moving in a given sense normally to a given wall. The pressure caused by molecular bombardment in this case follows as

$$p = \tfrac{1}{3}nmv^2 = \tfrac{2}{3}u_{\mathrm{M}} \tag{19.10}$$

where, in this case, $u_{\mathrm{M}} = \frac{1}{2}nmv^2$ is the energy density of the molecules within the enclosure.

A similar procedure can be applied to the case of radiation within an enclosure. Referring to Fig. 19.7, it is seen on recalling equation (19.7) that the energy density of radiation at element $\mathrm{d}A'$, which is located at the centre of the spherical cavity, is

$$\mathrm{d}u = \frac{K\,\mathrm{d}A}{cr^2}$$

Hence the net force on the upper surface of the element $\mathrm{d}A'$ is given by

$$\mathrm{d}^2F = \frac{K\,\mathrm{d}A\,\mathrm{d}A'\cos\theta}{cr^2}$$

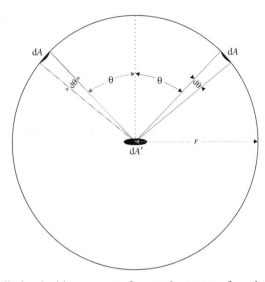

FIGURE 19.7 Radiation incident on a surface at the centre of a spherical cavity.

where the components of forces horizontal to the central surface element cancel because of symmetry and therefore only the components in the normal direction require consideration. Since $dA = 2\pi r^2 \sin\theta \, d\theta$ on the spherical surface, the total force exerted on the upper surface of dA by radiation from the inner area of the spherical cavity is

$$dF = \frac{2\pi r^2 \, dA' K}{cr^2} \int_0^{\pi/2} \sin\theta \cos\theta \, d\theta = \frac{2\pi \, dA' K}{c} \int_0^{\pi/2} \sin\theta \cos\theta \, d\theta$$

However, the pressure of radiation on dA' is only dependent on the vertical components of d^2F above the area dA'; since, because of symmetry, all horizontal components of force cancel:

$$p_A = \frac{dF}{dA'} = \frac{2\pi K}{c} \int_0^{\pi/2} \sin\theta \cos^2\theta \, d\theta = \frac{2\pi K}{c} \int_{\pi/2}^0 \cos^2\theta \, d(\cos\theta)$$

an expression which readily reduces to

$$p_A = \frac{1}{3} \frac{2\pi K}{c} \left[\cos^3\theta\right]_{\pi/2}^0 = \frac{2\pi K}{3c} \tag{19.11}$$

At this stage, it is necessary to consider the emission of radiation from the central element dA' itself. In fact, if it is a black body in equilibrium within the cavity, it must be radiating as much energy as it receives. If, on the other hand, dA' is a full reflector, the energy density over the element is still doubled by the latter's presence. It is thus possible to argue inductively that the radiative energy density would be doubled by virtue of the presence of any body in the cavity, no matter what its radiative properties. It can thus be concluded that the actual pressure of radiation is double that given by equation (19.11), so that

$$p = \frac{1}{3} \frac{4\pi K}{c} = \frac{1}{3} u \tag{19.12}$$

where the right hand side of (19.12) follows from (19.9).

A factor of two has changed the expression for the pressure in (19.12) from that given for the molecular case in (19.10). In case this causes bewilderment, it might also be recalled that while the energy of a classical particle or molecule is $\frac{1}{2}mv^2$, that of a photon is mv^2 (where in the latter case $v = c$), also differing by a factor of two.

19.2 THERMODYNAMIC CYCLES

This discussion of thermodynamic cycles will be restricted to reversible processes, which are, in simplest terms, ones in which the pressure corresponding to a given volume of working substance remains constant. Although it may seem strange to refer to radiation as a working substance, it must be recalled that in the development of thermodynamic theory so far, apart from the concept of velocity of radiation, it has not been necessary to assume any particular properties which did not follow from fundamental and simple observations. Just as a gas was regarded as a working substance in the kinetic theory, and able to do work when allowed to expand the volume of a cylinder through exerting pressure on a movable piston, there is no reason why radiation, which has also been shown to exert a pressure and have a temperature-dependent energy density, could not be regarded as a working substance in a suitably designed cavity.

The simplest cycle is that discovered by Carnot, in which a working substance (which does not require specification) is followed through a sequence of adiabatic and isothermal changes to complete a closed path on a p–V diagram, illustrated in Fig. 19.8. The reason for choosing these two types of thermodynamic change is that well-understood mathematical expressions, most simply derived for an ideal gas, are available to describe them.

The ideal gas equation, and in particular its differential form, offers a logical starting point for the purpose of discussing the cycle:

$$pV = RT$$
$$p\,dV + v\,dp = R\,dT \tag{19.13}$$

When a gas or working substance is allowed to expand and do work as a result of the absorption of heat dQ, part of this energy increases the internal energy of the system and the balance is converted to external work:

$$dQ = c_v\,dT + p\,dv \tag{19.14}$$

For an adiabatic change, $dQ = 0$, so that from (19.14) and (19.13) it follows that

$$dT = -\frac{1}{c_v}(p\,dv)$$

which on substitution in (19.13) yields

$$p\,dV + V\,dp + \frac{R}{c_v}p\,dV = 0$$

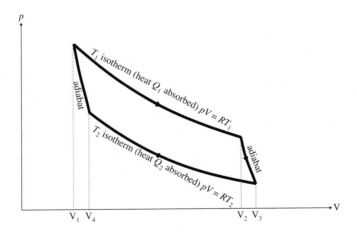

Figure 19.8 The four stages of the Carnot cycle.

Therefore

$$p \, dV \left(\frac{c_v + R}{c_v} \right) = \frac{c_p}{c_v} p \, dV = V \, dp \qquad (19.15)$$

and hence

$$\frac{dp}{p} = \frac{dV}{V} \frac{c_p}{c_v}$$

where $c_p = (c_v + R)$ is the specific heat at constant pressure. If $\gamma = c_p/c_v$ then the conditions for adiabatic change follow as either:

$$pV^\gamma = \text{constant} \qquad (19.16)$$

or, on substitution for p by T, from the first of the equations (19.13):

$$TV^{\gamma-1} = (\text{another}) \, \text{constant} \qquad (19.17)$$

The equation describing an isothermal change is simply an expression of Boyle's law, or alternatively equation (19.13) for a constant value of T:

$$pV = \text{constant} \qquad (19.18)$$

If a reversible, or frictionless, cycle as illustrated in Fig. 19.8 is now considered to commence with an isothermal change during which heat Q_1 is added to the working substance initially occupying a volume V_1 and V_2 at the end of this first phase of the cycle, then

$$Q_1 = \int_{V_1}^{V_2} dQ = \int_{V_1}^{V_2} p \, dV$$

and since $pV = RT$, it is clear that:

$$Q_1 = \int_{V_1}^{V_2} \frac{RT_1}{V} \, dV = RT_1 \ln \frac{V_2}{V_1} \qquad (19.19a)$$

and similarly for the second isothermal stage:

$$Q_2 = \int_{V_3}^{V_4} \frac{RT_2}{V} \, dV = RT_2 \ln \frac{V_4}{V_3} \qquad (19.19b)$$

The values of the ratios of the volumes in both equations (19.19) can be found after considering the consequences of the intermediate adiabatic changes for which equation (19.17) gives

$$T_1 V_2^\gamma = T_2 V_3^\gamma$$
$$T_2 V_4^\gamma = T_1 V_1^\gamma \qquad (19.20)$$

from which above pair of equations it follows that $V_2/V_1 = V_3/V_4$, which when substituted in the two equations (19.19) results in

$$\frac{Q_1}{T_1} = \frac{Q_2}{T_2}$$
$$\frac{(Q_1 - Q_2)}{Q_1} = \frac{(T_1 - T_2)}{T_1} \qquad (19.21)$$

Now recalling that if Q_1 is the heat absorbed by the system and Q_2 that given out, then the external work done is

$$w = Q_1 - Q_2$$

so that the efficiency of the process is given by $w/Q_1 = (Q_1 - Q_2)/Q_1$, a ratio which, together with (19.21), is fundamentally important.

The great significance of the Carnot cycle lies in the fact that any reversible thermodynamic change can be built up by a sequence of sufficiently small isothermal and adiabatic steps. The concept of this cycle also leads to the definition of a very useful thermodynamic term known as the entropy.

A substance undergoing reversible change and taking in heat dQ whilst at a temperature T is said to increase its entropy by:

$$dS = \frac{dQ}{T} \tag{19.22}$$

The change in entropy depends only on its initial and final conditions and not on the particular reversible process by which the working substance passes from one state to another.

In considering the entropy changes occurring during the four component steps of a Carnot cycle, it is evident that during the two separate adiabatic changes, no change in entropy occurs. However, during the two isothermal changes, entropy increases by Q_1/T_1 and decreases by Q_2/T_2, so that the net change in entropy is given by

$$\delta S = S_1 - S_2 = \frac{Q_1}{T_1} - \frac{Q_2}{T_2} \tag{19.23}$$

a quantity which from equation (19.21) is seen to be zero. Hence, *during any reversible cycle, entropy change is zero*, but it is important to note that the net change in heat energy is not zero when mechanical work has been done.

It should be noted that in using the term 'thermodynamic working substance', there has been no concern expressed in relation to the details of the physical nature of the substance. It will become clear that for most purposes it is irrelevant whether a cavity contains the ideal gas introduced in treatises of elementary thermodynamics, or radiation. It has not, as yet, been found essential to enquire into the actual nature of radiation, since all the basic knowledge has followed from the fact that radiation, at least up to the present stage of development of the theory, is simply a convenient term describing the process whereby thermal equilibrium can be achieved between bodies in a vacuum.

Following from the first law of thermodynamics (an expression of the law of conservation of energy), if a substance of volume V absorbs heat dQ and external forces do work dw on this substance, then its increase in internal energy is given by

$$dU = V \, du$$

$$= dQ + dw \tag{19.24}$$

(in which it is important to distinguish between the total internal energy U and the energy density u).

If the external force is observed as a uniform pressure, p, then

$$\mathrm{d}w = -p\,\mathrm{d}V \tag{19.25}$$

so that from (19.24) and (19.25) it is evident that

$$V\,\mathrm{d}u = \mathrm{d}Q - p\,\mathrm{d}V$$

and on introducing the expression for entropy (19.22), it follows that

$$V\,\mathrm{d}u = T\,\mathrm{d}S - p\,\mathrm{d}V \tag{19.26}$$

19.3 THE STEFAN–BOLTZMANN LAW

Measurements of the rate of cooling of a body as a function of its temperature, and that of its surroundings, were made by Dulong and Petit in 1819 and extended by Tyndall in 1865. The results of these investigations can be expressed in the form

$$\frac{\mathrm{d}T}{\mathrm{d}t} = f(T) - f(T_0) \tag{19.27}$$

where T is the temperature of the body and T_0 is that of its surroundings.

In 1879 Stefan showed that $f(T)$ is proportional to the fourth power of the temperature. These observations were later supported by the theoretical analysis of Boltzmann in 1884.

The Stefan–Boltzmann law states that if a black body at absolute temperature T is surrounded by another black body at absolute temperature T_0, then the net rate of transfer of radiative energy at the former surface is given by:

$$E = \sigma(T^4 - T_0^4) \tag{19.28}$$

where σ is known as Stefan's constant. In this expression, if $T_0 = 0$, then E is identical to the emissive power of the black body at temperature T.

As originally shown by Boltzmann, this law can be established on purely thermodynamic grounds. This demonstration makes use of the thermodynamic parameters which have been discussed earlier and relies on recalling that the energy density of radiation, u, in an enclosure depends only on the temperature, $u = u(T)$, and that the pressure on a surface element in such an enclosure is given by $p(V, T) = \frac{1}{3}u(T)$.

Because of the known relationship between pressure, volume and temperature, the concept of a reversible cycle, which was introduced earlier to explore the effect of a change in the volume of a radiation cavity, the energy density–temperature function, is again employed. In this instance, the behaviour of the pressure is initially examined.

If an infinitesimal quantity of heat $\mathrm{d}Q$ is led into a cavity, conveniently regarded to be in the form of a cylinder equipped with a piston, it is possible to examine the consequent changes in volume $\mathrm{d}V$, temperature $\mathrm{d}T$, pressure $\mathrm{d}p$ and energy density $\mathrm{d}u$. The change is regarded as being reversible, but certainly not adiabatic, because heat has been introduced, nor is it isothermal. The heat supplied causes an increase in internal energy as well as allowing the 'working substance', radiation, to do work by expansion:

$$dQ = d(uV) + p\,dV = V\,du + u\,dV + p\,dV$$

but because $p = \frac{1}{3}u$, it follows that

$$dQ = V\,du + \tfrac{4}{3}u\,dV \tag{19.29}$$

In order to specify the process by functions which are dependent on the condition of the system only, the quantity of heat, dQ, is replaced by a term containing the entropy, $T\,dS$, so that

$$dS = \frac{4}{3}\frac{u}{T}\,dV + \frac{V}{T}\,du \tag{19.30}$$

With dS being a perfect differential, it is possible to write

$$dS = \left(\frac{\partial S}{\partial V}\right)_u dV + \left(\frac{\partial S}{\partial u}\right)_V du$$

Hence

$$\left(\frac{\partial S}{\partial V}\right)_u dV = \frac{4}{3}\frac{u}{T} \quad \text{and} \quad \left(\frac{\partial S}{\partial u}\right)_V du = \frac{V}{T}$$

Since

$$\left(\frac{\partial}{\partial u}\right)_V \left(\frac{\partial S}{\partial V}\right)_u = \left(\frac{\partial}{\partial V}\right)_u \left(\frac{\partial S}{\partial u}\right)_V$$

it follows that

$$\left(\frac{\partial}{\partial u}\right)_V \left(\frac{4u}{3T}\right) = \left(\frac{\partial}{\partial V}\right)_u \left(\frac{V}{T}\right)$$

But since u is a function of T only

$$-\frac{4u}{3}\frac{1}{T^2}\left(\frac{\partial T}{\partial u}\right)_V + \frac{4}{3T} = -V\frac{1}{T^2}\left(\frac{\partial}{\partial V}\right)_u + \frac{1}{T}$$

$$\left(\frac{\partial T}{\partial u}\right)_V = \frac{dT}{du} \quad \text{and} \quad \left(\frac{\partial T}{\partial V}\right)_u = 0$$

Therefore

$$\frac{4}{3}\frac{u}{T^2}\frac{dT}{du} = \frac{1}{3T}$$

and thus

$$4\frac{dT}{T} = \frac{du}{u} \tag{19.31}$$

Integration of equation (19.31) leads to

$$u(T) = aT^4 \tag{19.32}$$

where a is a constant of integration. This equation can be expressed in terms of the emissive power, by substituting equation (19.19)

$$E = \frac{ac}{4} T^4 = \sigma T^4 \qquad (19.33)$$

Various experimental determinations of the value of Stefan's constant, σ, have indicated that

$$\sigma = 5.67032 \times 10^{-8} \, \text{W} \, \text{m}^{-2} \, \text{K}^{-4}$$

As everyday examples, simple evaluations of equation (19.33) reveal that $10 \, \text{mm}^2$ of a perfectly black body at 3000 K, simulating an electric light filament, would emit radiant energy at a rate of about 46 W, while a cylindrical electrical heating element at a temperature of 1000 K having a surface area of $10\,000 \, \text{mm}^2$ would radiate about 570 W.

It has now been seen that the relatively simple concept of the radiation cavity allowed the methods of classical physics to explain the relationship between total radiated power and the temperature of the radiating surface. Indeed the explanation of Stefan's empirical law was a major scientific triumph. However, a classical physical explanation for the fundamental observation of the change of colour of radiators with temperature remained elusive.

As long as radiation is restricted to the visible range, simple concepts of a black body absorber can be readily visualized in the form of either a matt black painted surface, or even as an actual cavity, with a small entrance which can be 'seen' as a very efficient black surface. However, other ranges of radiation must be associated with other types of absorbing surfaces. An extreme example at the higher frequency end of the known spectrum is provided by nuclear radiations, such as gamma rays, to which the types of black body which have been considered so far may be quite transparent. At relatively low frequencies, compared with thermal and visible radiation, radio waves serve as an example. The latter are not affected by the colour of intercepting surfaces at all.

As the nineteenth century drew to a close, physicists realized that simply considering radiation as an unspecified 'working substance' would not lead to further advances. New assumptions had to be made and tested.

In the discussion of the thermodynamics of radiation so far, no discussion of the actual nature of radiation was found to be necessary. However, with the development of 'grating' spectroscopy in the nineteenth century, the wavelength characteristics of not only visible light but also radiation perceived as heat were investigated. The latter were discovered to lie in a range of wavelengths just longer than the visible red and thus became known as infrared. Another group of wavelengths were identified as being just shorter than those of the visible violet and were consequently designated as ultraviolet radiation.

19.4 THE BLACK BODY SPECTRUM AND WIEN'S DISPLACEMENT LAW

Following the discovery that any body emits radiation over a range of wavelengths, it was realized that whereas the Stefan–Boltzmann law gives the total power of radiation of all wavelengths emitted, E, as a function of the absolute temperature, it became necessary to introduce a modified notation such that $dE(\lambda)$ is the rate of radiation having wavelengths between λ and $\lambda + d\lambda$ from unit area of a radiator. Since the quantity $dE(\lambda)$ is infinitesimal in magnitude, a new parameter is defined such that

$$E_\lambda = \frac{dE(\lambda)}{d\lambda} \tag{19.34}$$

and the plot of E_λ against λ is said to represent the spectrum of the radiation.

The effort to specify the shape of the spectrum of black body radiation or the determination of the functional relationship between E_λ and the temperature became a great challenge for nineteenth-century physicists.

A readily made fundamental observation lies in the fact that as a suitable body is heated, e.g. electrically, initially invisible radiated heat may be felt. Further heating results in a red glow. Continuing the heating by applying more energy results in the light becoming more yellow and later even tinged with blue.

A quantitative spectrometer allows the energy in standard incremental wavelength ranges to be found. When this is done for various temperatures, the change in shape of the spectrum as well as the movement of the position of the maximum can be studied. Figure 19.9 shows black body spectra for a range of emitter temperatures.

During the last decade of the nineteenth century, Wien made an intensive study of radiation spectra and observed that:

1. The values of $E_\lambda = dE/d\lambda$ increase as T increases and the Stefan–Boltzmann law is obeyed as follows:

$$\int_0^\infty \frac{dE}{d\lambda}\, d\lambda = \sigma T^4 \tag{19.35}$$

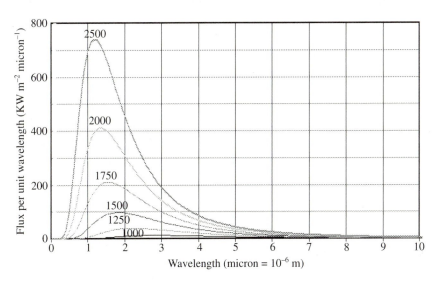

FIGURE 19.9 Black body radiation spectra for various temperatures shown in Kelvin.

2. With increasing temperature T, the position of the maximum value of E_λ shifts towards the shorter wavelengths. In 1893, Wien discovered that the exact way in which this maximum value shifts is given by

$$\lambda_m T = \text{constant} \qquad (19.36)$$

an equation which is known as Wien's displacement law and which was experimentally verified by Lummer and Pringsheim in 1899. Figure 19.10 has been constructed from the wavelengths for the maxima in the spectra shown in Fig. 19.9.

Wien's observation on the position of the maximum of spectra and subsequent theoretical explanation became the last major contribution of the methods of classical physics to the understanding of radiation.

In his theoretical analysis, Wien again relied on considering the conditions which describe radiation in an adiabatically expanding cavity. As such, this procedure could be described as being similar to that used in establishing the Stefan–Boltzmann law. The main innovation was to consider radiation to be in the form of standing waves, whose wavelengths were determined by cavity dimensions.

If radiation in equilibrium with its enclosure is visualized as being constituted of standing waves, then if a characteristic linear dimension of the cavity is increased from l_1 to l_2, the wavelength of the associated standing-wave radiation must change from λ_1 to λ_2 in accordance with the following relation:

$$\frac{\lambda_1}{\lambda_2} = \frac{l_1}{l_2} \qquad (19.37)$$

This theory is compatible with the observation that expansion reduces the energy density of cavity radiation and also the equilibrium temperature as

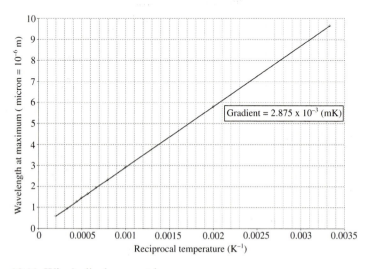

FIGURE 19.10 Wien's displacement law.

well as the consequent inference that the characteristic wavelength, or that of the maximum, associated with the cavity would increase.

Whereas it was earlier demonstrated that the energy density of radiation depends only on the temperature and not on a cavity's shape, Wien's theoretical development relies on assumptions as to a cavity's geometry. In particular, it is necessary to consider that any adiabatic change in the dimensions of a cavity need to be of such a nature as to ensure the maintenance of the relativity between the continuous range of all wavelengths which the cavity supports.

In order to discuss this type of phenomenon on a precise basis, attention is restricted to radiation within a small range of wavelengths. This necessitates introducing the concept of the spectrum of any relevant property of radiation, not only that of the radiant flux. Typically, such parameters would include the energy density and the pressure of radiation:

$$u_\lambda = \frac{du(\lambda)}{d\lambda} \quad \text{and} \quad p_\lambda = \frac{dp(\lambda)}{d\lambda}$$

Attention is now directed to the energy density associated with an infinitesimal band of radiation in the initial state of a cavity before an adiabatic expansion, $u_{\lambda_1} d\lambda_1$, and after expansion, $u_{\lambda_2} d\lambda_2$, where it is defined that

$$\frac{\lambda_1}{\lambda_2} = \frac{d\lambda_1}{d\lambda_2} \tag{19.38}$$

The argument that led to the theoretical formulation of the Stefan–Boltzmann law for total, or all wavelengths of, radiation is now applied to an infinitesimal band of wavelengths, for which the equation describing adiabatic change becomes

$$d(u_\lambda \, d\lambda V) + p_\lambda \, d\lambda \, dV = 0 \tag{19.39}$$

where $p_\lambda \, d\lambda$ represents that part of the pressure caused by the wavelengths under consideration.

From equation (19.12) it now follows that

$$p_\lambda \, d\lambda = \tfrac{1}{3} u_\lambda \, d\lambda \tag{19.40}$$

and thus on substituting (19.40) into (19.39)

$$d(3V(p_\lambda \, d\lambda)) + (p_\lambda \, d\lambda) \, dV = 0$$

Therefore

$$3V \, d(p_\lambda \, d\lambda) + 4(p_\lambda \, d\lambda) \, dV = 0$$

$$\frac{d(p_\lambda \, d\lambda)}{(p_\lambda \, d\lambda)} + \frac{4}{3} \frac{dV}{V} = 0$$

$$\ln(p_\lambda \, d\lambda) + \tfrac{4}{3} \ln V = \ln(p_\lambda) V^{4/3} = \text{constant}$$

from which it then follows that the adiabatic expansion of the cavity is described by

$$p_{\lambda_1} \, d\lambda_1 V_1^{4/3} = p_{\lambda_2} \, d\lambda_2 V_2^{4/3} \tag{19.41}$$

If equation (19.41) is rewritten in terms of the energy density u and the characteristic linear cavity dimension l then assuming that the cavity volume V is proportional to l^3 the expression becomes

$$u_{\lambda_1}\, d\lambda_1 l_1^4 = u_{\lambda_2}\, d\lambda_2 l_2^4 \tag{19.42}$$

but through application of equations (19.37) and (19.38), it is possible to remove the infinitesimal quantities $d\lambda$ and the somewhat artificial concept of characteristic cavity dimension l from (19.42) to yield

$$\frac{u_{\lambda_1}}{u_{\lambda_2}} = \frac{\lambda_2^5}{\lambda_1^5} \quad \text{or} \quad u_{\lambda_1}\lambda_1^5 = u_{\lambda_2}\lambda_2^5 \tag{19.43}$$

While equation (19.43) is one which is characteristic of the adiabatic cavity expansion, it might also be claimed that the Stefan–Boltzmann equation itself (19.32), if rewritten for the same infinitesimal band of wavelengths, would also be a characterizing expression. If this is done, then

$$\frac{u_{\lambda_1}\, d\lambda_1}{u_{\lambda_2}\, d\lambda_2} = \frac{T_1^4}{T_2^4} = \frac{u_{\lambda_1}\lambda_1}{u_{\lambda_2}\lambda_2} \tag{19.44}$$

Equations (19.43) and (19.44) are two independent expressions, but it is possible to generate a third equation by eliminating the u_λ ratio as follows:

$$\lambda_1^4 T_1^4 = \lambda_2^4 T_2^4 \quad \text{or} \quad \lambda_1 T_1 = \lambda_2 T_2 \tag{19.45}$$

It should be noted that while the second of the above equations is the simpler, it is clear that in general the equality would apply to any function of λT as shown below:

$$f(\lambda_1 T_1) = f(\lambda_2 T_2) \tag{19.46}$$

There are now three expressions, from which a selection can be made to provide two independent equations to combine to specify the adiabatic expansion of the cavity. From these, (19.43) and (19.46) are chosen to produce

$$\frac{u_{\lambda_1}\lambda_1^5}{f(\lambda_1 T_1)} = \frac{u_{\lambda_2}\lambda_2^5}{f(\lambda_2 T_2)} \tag{19.47}$$

Equation (19.47) implies that

$$u_\lambda = \frac{A}{\lambda^5} f(\lambda T) \tag{19.48}$$

where A is a constant.

Equation (19.48) has been derived on the basis of thermodynamic reasoning alone, apart from the necessary visualization of standing waves in a radiation cavity. Although the methods of classical physics were not able to determine the nature of the function $f(\lambda T)$ considerable progress was achieved in describing the spectrum of black body radiation.

Occasionally, it is convenient to discuss the spectrum in terms of the frequency of radiation v so that the concept of energy density per unit frequency interval is introduced as $u_v\, dv$ wherein it is defined that

$$u_v\, dv = u_\lambda\, d\lambda$$

but since $\lambda = c/\nu$, where c is the velocity of radiation,

$$d\lambda = -c\,\frac{d\nu}{\nu^2}$$

$$\tag{19.49}$$

$$u_\nu = B\nu^3 g(\nu/T)$$

where B is another constant (such that $B = A/c^4$) and $g(\nu/T) = f(\lambda T)$.

Wien's displacement law can be shown to follow as a particular consequence of equation (19.48) for the expression for u_λ when this function is itself considered as a function of the temperature. From equation (19.45) it is clear that $1/T^5$ can be substituted for λ^5 in (19.48) to yield

$$u_\lambda/T^5 = Af(\lambda T) \tag{19.50}$$

If (19.50) is viewed as a simple functional relationship between two variables, say x and y, so that $y = Af(x)$, it would be reasonable to anticipate that a plot of y against x would be a single curve. If this function has a maximum value $x_{max} = \lambda_{max} T$ then this value would be the same for all values of T, i.e. $\lambda_{max} T = $ constant.

19.5 WIEN'S EXPRESSION FOR THE FREQUENCY DISTRIBUTION OF RADIATION

Following his success with the displacement law, Wien struggled in vain to offer a theoretical explanation for the observed shape of the black body spectrum. Finally he resorted to the creation of a semi-empirical expression of the following form:

$$u_\lambda = c_1 \lambda^{-5}\, e^{-c_2/\lambda T} \tag{19.51}$$

in which equation the two constants c_1 and c_2 required empirical determination.

While the presence of the λ^{-5} term on the right of equation (19.51) follows logically from (19.48), the exponential part of the expression appears to be an attempt to guess the nature of the function $f(\lambda T)$. Nevertheless, Wien's equation does give $u_\lambda = 0$ for $\lambda = 0$ and has a single maximum value as required. Wien was able to select values for the two constants c_1 and c_2 to force (19.51) to fit experimental observations at wavelengths shorter than those associated with spectral maxima, only to be unable to correct for small but significant departures from reality at longer wavelengths. As will be shown later, with the above optimal selection of values for the constants, spectral integration leads to an underestimation of the total radiation by about 8%, which for many purposes represents a not too unsatisfactory result by this final classical physical contribution.

19.6 OSCILLATORS, RADIATORS AND SPECTRA

The impasse reached by the methods of classical thermodynamics became partially resolved by the advent of James Maxwell's electromagnetic theory, which in demonstrating the connection between the properties of light and the nature of electricity and magnetism was one of the great contributions in nineteenth-century mathematical physics. In this theory, the electrons in neutral matter are linked with the processes of electrical and thermal conduction. As a further step, radiation is considered to be in the form of electromagnetic waves generated by oscillating electrons. In this context, the problem of elucidating cavity radiation became transformed into the study of the equilibrium between a large collection of oscillators which could absorb as well as emit radiative energy.

A particular electron of charge e and mass m is considered to be undergoing simple harmonic motion with angular frequency $\omega_0 = 2\pi\nu_0$, and if the mechanical energy of this electron is given by ε then Maxwell showed that the rate of radiation is given by

$$E_{\nu_0} = \frac{8\pi^2 e^2 \nu_0^2}{3mc^3}\varepsilon \qquad (19.52)$$

where c is the velocity of light.

The same oscillating electron is now considered to be exposed to a plane wave of radiation. Maxwell was also able to show that the total power absorbed is concentrated in a very narrow band situated in the immediate vicinity of the electron's natural frequency of oscillation ν_0 and if u_{ν_0} is the value of the energy density spectrum at the resonant frequency ν then the power absorbed is given by

$$E_{\nu_0} = \frac{\pi e^2}{3m}u_{\nu_0} \qquad (19.53)$$

In order for an electron to achieve equilibrium, the rates of emission and absorption must become equal. Rather than remain with only a single electron as in the above two sections, a large collection of electronic oscillators is now considered, so that the possibility exists for any electron to have any arbitrary frequency, and hence energy. This means that the equilibrium condition for a sufficiently large group of oscillators does not need to be restricted to a specific frequency ν_0 but to a range of frequencies ν and is therefore given by

$$\frac{8\pi e^2 \nu^2}{3mc^3}\bar{\varepsilon} = \frac{\pi e^2}{3m}u_\nu$$

$$u_\nu = \frac{8\pi\nu^2}{c^3}\bar{\varepsilon} \qquad (19.54)$$

This equation connects the energy density of radiation with the mean mechanical energy $\bar{\varepsilon}$ of a system of oscillators, where, because the expression refers to a group of oscillators rather than a single one, it remains necessary to determine an expression for $\bar{\varepsilon}$ in order to be able to specify the whole spectrum of radiation.

This theory attempted to link a theory of radiation to concepts of the then recently developed kinetic theory of gases. The likely inspiration for this lay in the earlier classical thermodynamic analyses in which radiation was regarded as a working substance behaving similarly to a gas. One of the important contributions of the kinetic theory was to provide a relationship between the energy of gaseous molecules and the absolute temperature. Boltzmann and

Maxwell in their development of statistical mechanics showed that a far-reaching principle, that of the equipartition of energy, governed the distribution of energy amongst the various modes of motion open to a molecule. This principle states that a mean energy of $\frac{1}{2}kT$ is associated with each degree of freedom, where T is the absolute temperature and k is a constant now known as the Boltzmann constant.

The English mathematical physicists, Rayleigh first and Jeans later, applied this theory of energy partition to an assembly of oscillating electrons. In this model, the electrons were assumed to be fixed in mean positions with only two degrees of vibrational freedom, so that the mean total energy of an oscillator of the type considered as radiators and oscillators in sections 3.1, 3.2 and 3.3 would involve two multiples of $\frac{1}{2}kT$ and hence

$$\bar{\varepsilon} = kT \tag{19.55}$$

a simple expression which can readily be applied to equation (19.51) assumedly to yield an equation for the spectrum of black body radiation as follows

$$u_\nu = \frac{8\pi\nu^2}{c^3}kT \tag{19.56}$$

This result was obtained by Rayleigh in 1900 and by Jeans in 1905 and was found to provide a good fit to the experimental results in the region of long wavelengths, for which of course ν has small values. The crucial objection is almost obvious from an inspection of equation (19.53) and became historically famous as the ultraviolet catastrophe. As the wavelength tends to zero, the frequency tends to infinity, so that the Rayleigh–Jeans model indicates that the energy density should also tend to infinity. This is, of course, nonsense, even if only because integration of the energy density over all frequencies leads to infinite values of the total energy density at any non-zero temperature. This rigorous adhesion to classical principles had generated a blatant falsehood.

19.7 PLANCK'S QUANTUM THEORY

The failure of classical physics as first demonstrated by Rayleigh led the German physicist Max Planck to re-examine the subject of radiation from fundamental principles in 1900. Instead of allowing atomic oscillators to have an arbitrary energy, with a specification only of the total energy of an assembly of oscillators, Planck postulated that they were restricted to integral multiples of some basic unit.

Although the photo-electric effect which exhibited a response threshold dependent on the frequency of light was a strongly supporting observation, at the time, Planck's concept was revolutionary, so much so that Jeans, in apparent disbelief, persevered on classical lines in his attempts to elucidate the problem as late as the year 1905.

In suggesting that the energy of oscillating electrons should be regarded as restricted to integral multiples of some basic unit ε_0 with a restriction on the proportion of the total number of electrons able to occupy a particular level of energy, Planck departed radically from the classical case in which the Boltzmann distribution function appropriately describes the continuous range of energies which molecules may have in the kinetic theory. In the latter, the number of molecules which have an energy between ε and $d\varepsilon$ is given by

$$dn(\varepsilon) = A\,e^{-\varepsilon/kT}\,d\varepsilon \tag{19.57}$$

where A is a constant. Planck's discontinuous, permissible energies are shown in Table 19.1.

Planck consequently was able to show that the unconstrained term for the mean energy $\bar{\varepsilon}$ in equation (19.54) should be replaced by

$$\bar{\varepsilon} = \frac{\varepsilon_0}{e^{\beta\varepsilon_0} - 1} \tag{19.58}$$

where β is a constant, analogous to one encountered in Maxwell–Boltzmann statistical mechanics, the magnitude of which, in the present instance determines the spacing of the energy levels of the electrons prescribed by Planck.

The expression for the frequency distribution, or spectrum, of black body radiation thus becomes

$$u_\nu = \frac{8\pi\nu^2}{c^3}\,\frac{\varepsilon_0}{e^{\beta\varepsilon_0} - 1} \tag{19.59}$$

It is now interesting to explore the link between the Planck black body spectrum and those of the classically based theories.

Examining the exponential part of equation (19.59) by noting that

$$e^x = 1 + \frac{x}{1!} + \frac{x^2}{2!} + \frac{x^3}{3!} + \cdots$$

and if $x = \beta\varepsilon_0$ is much less than 1, then

$$e^{\beta\varepsilon_0} - 1 \to \beta\varepsilon_0$$

Thus if β is considered to tend to a value much less than 1 in equation (19.59), it becomes evident that the separation between the permitted energies of oscillation would also tend to zero, since then

$$u_\nu = \frac{8\pi\nu^2}{c^3}\,\frac{1}{\beta} \tag{19.60}$$

It should be observed that if $1/\beta = kT$ in equation (19.60), then the Rayleigh–Jeans formula for the spectrum (19.55) is reproduced.

It is now interesting to compare the spectral equation (19.59) derived by

Table 19.1 Numbers of electron oscillators at specified energy levels in Planck's theory (A is a constant)

Level of electron oscillator energy	No. of electrons with this energy
$0 \times \epsilon_0 = 0$	$A = A\,e^0$
ϵ_0	$A\,e^{-\beta\epsilon_0}$
$2\epsilon_0$	$A\,e^{-2\beta\epsilon_0}$
$n\epsilon_0$	$A\,e^{-n\beta\epsilon_0}$

Planck with the earlier thermodynamic expression (19.50) obtained by Wien, both of which are in the form showing the frequency distribution of radiation. As concluded earlier, although the Wien equation does not reveal the complete functional nature of the spectrum, the details offered are true because no unreasonable assumptions were made in the process of its derivation. Thus it is possible to write

$$B\nu^3 g\left(\frac{\nu}{T}\right) = \frac{8\pi\nu^2}{c^3} \frac{\varepsilon_0}{e^{\beta\varepsilon_0} - 1} = \frac{8\pi\nu^3}{c^3} \frac{\varepsilon_0/\nu}{e^{\beta\varepsilon_0} - 1} \tag{19.61}$$

If now the substitution $\beta = 1/kT$ is made in (19.61) so as to make it compatible with the Rayleigh–Jeans formula when the energy level spacing $\varepsilon_0 \to 0$, i.e. a continuous range of energies is permitted for the electrons, then for (19.61) to be true in general, it must be that

$$B = \frac{8\pi}{c^3}$$

and

$$g\left(\frac{\nu}{T}\right) = \frac{\varepsilon_0/\nu}{e^{\varepsilon_0/kT} - 1}$$

The second of these above conditions can only be met if $\varepsilon_0/kT \propto \nu/T$ and hence $\varepsilon_0 \propto \nu$. The constant of proportionality between ε_0 and ν is given the symbol h and named Planck's constant.

The energy density spectrum of black body radiation can now be written as

$$u_\nu = \frac{8\pi\nu^3}{c^3} \frac{1}{e^{h\nu/kT} - 1} \tag{19.62}$$

This equation is known as Planck's formula and was experimentally shown by Rubens and Michel in 1919 to represent the black body spectrum accurately from $-160°C$ to $1800°C$. Over this range of temperatures, it is significant to observe that the Stefan–Boltzmann law indicates that the total radiant power changes by a factor of about 2.5×10^5 which shows how well the Planck formula describes this phenomenon.

When expressed in terms of wavelengths, the Planck formula follows from the application of a substitution of c/ν for λ, similar to that used in generating equation (19.48), the classical expression for u_ν, from (19.49), the corresponding one for u_λ, as shown below

$$u_\lambda = \frac{8\pi hc}{\lambda^5} \frac{1}{e^{hc/k\lambda T} - 1} \tag{19.63}$$

In order to generate an expression for the spectral emissive power E_ν or E_λ, it is simply necessary to recall, from equation (19.9), that $E = \frac{1}{4}uc$ where c, as throughout this text, is the velocity of light, e.g.

$$E_\lambda = \frac{2\pi hc^2}{\lambda^5} \frac{1}{e^{hc/k\lambda T} - 1} \tag{19.64}$$

Figures 19.9 and 19.10 were derived from values calculated using the above equation.

19.8 RELATIONSHIP BETWEEN THE STEFAN–BOLTZMANN, WIEN AND PLANCK LAWS

An expression similar to equation (19.34) allows the Stefan–Boltzmann law to be written in a form

$$\sigma T^4 = \frac{c}{4} \int_0^\infty u_\nu \, d\nu \tag{19.65}$$

which, when substituting the appropriate expression for u_ν offered by the Planck formula (19.62), gives the Stefan–Boltzmann constant as

$$\sigma = \frac{1}{T^4} \frac{8\pi h}{4c^2} \int_0^\infty \frac{\nu^3 \, d\nu}{e^{h\nu/kT} - 1} \tag{19.66}$$

If this expression is simplified by substituting $x - h\nu/kT$ the further substitutions follow:

$$d\nu = \frac{kT}{h} \, dx \quad \text{and} \quad \nu^3 = \frac{k^3 T^3}{h^3} x^3$$

so that:

$$\sigma = \frac{2\pi k^4}{h^3 c^2} \int_0^\infty \frac{x^3 \, dx}{e^x - 1} \tag{19.67}$$

Fortunately the integral on the right hand side of (19.67) is a standard one and in fact equal to $\pi^4/15$ so that

$$\sigma = \frac{2\pi k^4}{h^3 c^2} \frac{\pi^4}{15} \tag{19.68}$$

It is interesting to attempt to solve equation (19.67) by means of an approximation in which it is noted that the integrand tends to zero as $x \to 0$, while for all values of $x > 1$, $e^x \gg 1$, so that by neglecting the small inaccuracies arising through the approximation in the integration for values of x close to zero, the following solution results:

$$\sigma = \frac{2\pi k^4}{h^3 c^2} \int_0^\infty x^3 e^{-x} \, dx = \frac{2\pi k^4}{h^3 c^2} \times 6 \tag{19.69}$$

This expression could equally well have been generated by Wien's formula for the spectrum (19.51), which when rewritten for frequencies, rather than wavelengths, can be shown to become

$$u_\nu = \frac{b_1 \nu^3}{c^5} e^{-b_2 \nu/cT} \tag{19.70}$$

where b_1 and b_2 are another pair of Wien's positive, empirically determined, constants, related to those shown in (19.51). If the substitution $x = b_2 \nu/cT$ is made similarly to that leading to equation (19.67), then first

$$u_\nu \, d\nu = \frac{b_1 T^4 x^3}{b_2^5 c} e^{-x} \, dx$$

and subsequently by the same method which led to the expression for the Stefan–Boltzmann constant in equation (19.66)

$$\sigma = \frac{b_1}{4b_2^5} \int_0^\infty x^3 e^{-x} \, dx \tag{19.71}$$

There is an obvious similarity between equations (19.71), above, and (19.68) which implies that by appropriate choice of the Wien constants, the ratio between the spectrally integrated values of the Wien and Planck formulae is $6:\pi^4/15$, i.e. approximately 6:6.49, implying an error of about 8% only.

Returning to Planck's law expressed in terms of wavelengths, equation (19.63), and again simplifying the expression by making the substitution $x = hc/\lambda kT$, this becomes

$$u_\lambda = \frac{8\pi hc}{\lambda^5} \frac{x^5}{e^x - 1} \tag{19.72}$$

Planck's solution for the constant in Wien's displacement law can now be obtained simply through determining the value of x which is proportional to $1/\lambda T$ for which $du_\lambda/d\lambda = 0$. Determination of the appropriate values of x to achieve the above condition cannot be performed analytically, but a numerical solution can be obtained. In fact the value of $x = hc/\lambda kT$ for which $du_\lambda/d\lambda = 0$ is found to be 4.965.

Again it is interesting to determine the result through use of Wien's formula (19.51) which, with the substitution of $x = c_2/\lambda T$, is written as

$$u_\lambda = \frac{c_1}{c_2^5} T^5 x^5 e^{-x} \tag{19.73}$$

Determination of a stationary value of u_λ in equation (19.70) is simple and follows after differentiation

$$0 = 5x^4 e^{-x} - x^5 e^{-x}$$

from which it is clear that $x = 5.000$ which again shows how surprisingly well Wien's formula performs.

Again returning to the exact value of $x = hc/\lambda kT$, i.e. 4.965, for the maximum value of u_λ Planck's statement of the Wien displacement law would be

$$\lambda_{\max} T = \frac{hc}{4.965k} \tag{19.74}$$

Since the Stefan constant has been determined experimentally with considerable accuracy, and as listed in Section 19.3, $\sigma = 5.67 \times 10^{-8} \, \mathrm{W \, m^{-2} \, K^{-4}}$ so that from (19.68) it follows that

$$5.67 \times 10^{-5} = \frac{2\pi k^4}{h^3 c^3} \frac{\pi^4}{15} \tag{19.75}$$

Furthermore, the value of $\lambda_{\max} T = 0.29$ was reliably determined through the experiment of Lummer and Pringsheim, so that equation (19.74) becomes

$$0.29 = \frac{hc}{4.965k} \tag{19.76}$$

Finally, utilizing the fact that the velocity of light is given by $c = 2.998 \times 10^8 \, \mathrm{m\,s^{-1}}$, equations (19.75) and (19.76) can be solved for the hitherto unknown constants h and k which were introduced by Planck. When this is done, it is found that the constant k used by Planck takes an identical value to that of the Boltzmann constant which was deduced from kinetic theory. In fact

$$k = 1.38 \times 10^{-23} \, \mathrm{J\,K^{-1}}$$

and also

$$h = 6.625 \times 10^{-34} \, \mathrm{J\,s}$$

The logical link between the classical and quantum approaches to the explanation of the black body spectrum has thus been well demonstrated and there is every reason to have confidence in drawing on the Planck formula for quantitative spectral information.

The two equations shown for the Planck black spectrum, (19.62) and (19.63), giving the energy density of radiation in terms of frequency and wavelength respectively, lend themselves to being readily plotted by standard computer spreadsheet techniques when the values for c, k and h, all given above, are substituted. Not only was such a speadsheet used to generate Figs 19.9 and 19.10, but as will be seen later in Chapter 20, these data are useful in drawing other important quantitative conclusions.

20

THE RADIATION BALANCE OF THE EARTH

20.1 RADIATION AT THE EARTH'S SURFACE

During clear-sky daylight hours, the energy transformed at the Earth's surface is largely supplied in the form of incoming solar radiation, S_i, part of which, S_o, is scattered or reflected back, in the visible band of wavelengths. The standard unit of measurement for instantaneous levels of radiation, and indeed other meteorological energy fluxes, is $W\,m^{-2}$, where W (for Watt) = $J\,s^{-1}$ (or joules per second). This implies that a meteorological energy flux is normally specified as a power, or rate of supplying of energy, and consequently a daily or hourly total is given in units of work, usually per unit area, i.e. $J\,m^{-2}$. The Sun sends radiant energy at a rate of about $1370\,W\,m^{-2}$ to the 'top of the atmosphere' of the earth, which for an arbitrary fixed location averages over a 12 hour day to about $38\,MJ\,m^{-2}$, a quantity which in the usual, but non-standard, 'commercial' units employed by electrical power utilities equates to a little over $10\,kWh$ for every square metre. Depending on the cloud cover and atmospheric transparency generally, as well as the latitude, a lesser flux reaches the earth's surface.

Although ultraviolet radiation, most of which reaching the Earth is emitted by the sun, has important, particularly biological, impacts, it does not significantly influence the magnitude of the terrestrial energy balance. The wavelengths of ultraviolet radiation are shorter than those of the visible part of solar radiation. Ozone in the upper levels of the atmosphere serves a vital purpose for life on Earth by absorbing most of the incident solar ultraviolet radiation; hence the concern with the integrity of this protective layer. The popularly known 'ozone hole' in the Earth's upper atmosphere, while having important consequences to life on Earth, is almost irrelevant to the question of the Earth's energy balance.

The third, meteorologically important, wavelength grouping of radiation, known as the infrared, has longer wavelengths than those of the visible range.

The polyatomic constituents of the atmosphere, particularly water vapour and carbon dioxide (CO_2), both selectively absorb and reradiate radiant energy in a broad range of infrared bands. Because the sources and sinks for CO_2 are, apart from areas of industrial or combustional concentration, such as forest fires, relatively diffuse and slow acting, this gas is well mixed in the atmosphere and its contribution to radiant transfer processes is not subject to the great variations of those of water vapour, a gas with concentrations which vary greatly in space and time. In the atmosphere, water vapour is the dominant source of long-wave infrared radiation, and greatly exceeds any contribution made by CO_2 and other active gases of lesser concentration, including methane, CH_4. Infrared radiation is almost entirely terrestrial in origin, both in the form of inward energy reaching the Earth's surface from the atmosphere and clouds above, L_i, and the outward energy radiated by the solid surface itself, $L_o \approx \sigma T_0^4$ where $\sigma = 5.67 \times 10^{-8} \, \mathrm{W \, m^{-2} \, K^{-4}}$ is the Stefan–Boltzmann constant, and T_0 is the absolute temperature of the radiating surface. Since the emissivity of terrestrial surfaces is usually in the range of 95% to 98%, the simple approximation for L_o written above is satisfactory for many purposes.

Except during polar summers, the solar, the short-wave irradiance or the visible radiation flux vanishes as a significant source of energy during the night. The radiometers used in meteorological studies do not respond to even the brightest full moon. On the other hand, the long-wave or infrared radiation fluxes continue day and night, since this radiation depends only on emitters being above the absolute zero of temperature. The radiation fluxes discussed above are shown diagrammatically in Fig. 20.1.

FIGURE 20.1 Solar (short-wave, S, and ultraviolet, U) and terrestrial (long-wave, L) radiation fluxes at an idealized, opaque earth surface.

20.2 NET RADIATION AND ALBEDO

The algebraic sum of all of the four meteorologically significant radiant fluxes, discussed above, is known as the net radiation, R_N, defined by the equation

$$R_N = S_i - S_o + L_i - L_o \qquad (20.1)$$

This convenient parameter essentially summarizes the total energy available for partition between all other energy fluxes which originate or are transformed at the earth's surface. It is for this reason that net radiation, which can be measured with a single instrument, is often regarded as the most important single item of data in meteorological energy balance studies.

In spite of the importance of the direct measurement of net radiation, there are situations in which it is either necessary or preferable to measure the four components in equation (20.1) separately. While instrumentation exists to measure separately short- and long-wavelength ranges of radiation, it is essential to be confident that there is a minimal degree of overlap between these two meteorologically important bands of radiation.

The Earth's surface does not absorb all of the incoming short-wave, solar radiation S_i and an important property of the surface known as the albedo, β, determines what fraction is reflected and scattered back and if S_o is the magnitude of the latter flux

$$\beta = S_o / S_i \qquad (20.2)$$

Albedos of natural surfaces range from up to 90% for freshly fallen snow, to as low as 3% for the ocean when the sun illuminates it from overhead. Dry grassland has an albedo of about 25% and forests usually scatter back 10% or less. Whereas the short-wave radiation scattered upwards by some planar natural surfaces approximates cosine law estimates, with snow and bare, dry soil offering good examples, the reflectivity of many structured surfaces depends on the angle of incidence of the illumination, principally that from the sun, but under overcast conditions the illumination tends to be isotropic, i.e. equal from all directions. Figure 20.2 illustrates in two dimensions the incidence of two separate rays on a dark terrestrial surface such as soil, on which is growing an array of lighter-coloured plants. For the vertically incident beam, the magnitude of the albedo is mainly influenced by that of the soil, whereas for the other, more acute angle shown, the beam falls only on the lighter-coloured plant stalks, which entirely mask the view of the ground at the angle of incidence shown. It is intuitively clear that, in spite of secondary and subsequent scatterings which occur, the observed albedos of the same three-dimensionally structured surface will be markedly different in the simulated two cases, with the consequence that, in general, diurnal variations in albedos are to be expected.

Because of the possibility of high levels of surface reflectivity, rather than scattering, the albedo of calm water surfaces, including the sea under appropriately calm meteorological conditions, is particularly sensitive to solar attitude and consequently also to the state of the water surface. In general, all textured or structured surfaces, including forests but also, on a larger scale as might be

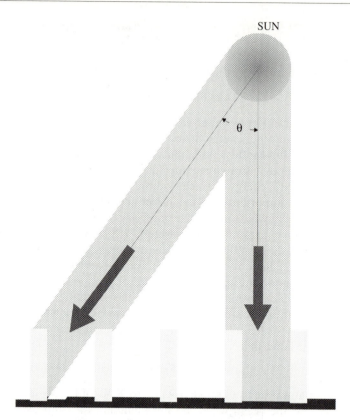

FIGURE 20.2 Illustration of the dependence of the albedo on the angle of incidence of radiation on three-dimensionally structured surfaces. The diagram simulates rows of light-coloured plant stalks growing on darker soil being irradiated, with beams shown from two different solar zenith angles.

viewed by satellite technology, landforms with significant topography, such as deserts with rows of dunes, have albedos which vary during the course of a day.

The concept of albedo is relatively straightforward when dealing with substantially uniform opaque surfaces, but becomes complex as soon as any significant level of translucency needs to be considered. Snow, clouds, sand and water surfaces clearly fall into this class. However, almost every natural surface, including those covered by vegetation, present optically complex structures for which the albedo cannot be described in terms of the properties of a single two-dimensional sheet.

The short-wave radiation returned from partially transparent or translucent terrestrial surface media is determined only in part by the properties of the surface, where the example of the reflection by water or by quartz crystals in sand comes to mind. But these same examples serve to emphasize the fact that incident radiation actually penetrates the structure of the material below the surface and because of the internal scattering processes, the flux of upwardly scattered radiation, in the case of translucency, will have been contributed to from a continuous range of depths.

20.3 THE FLUXES OF SOLAR AND TERRESTRIAL RADIATION

At the Earth's surface, the sun is seen as a source of radiation with an extent of about 0.68×10^{-4} steradians and a temperature of about 6000 K. The direct beam from this source, together with a diffuse component of a magnitude which depends on cloudiness and also clear sky scattering, constitutes the incoming short-wave radiation flux S_i. The spectral content of the short-wave radiation scattered and reflected by the Earth, S_o, depends on its optical properties at and near the surface, but the range of wavelengths cannot extend beyond the limits of that initially received from the Sun. The albedo of most terrestrial surfaces results in a substantially diffuse outward short-wave flux, although a direct beam or reflective component may be observable in many instances, particularly on still water surfaces, where reflectivity of the direct solar beam is dependent on the angle of incidence and may be the dominant signal. Whether the short-wave flux S_o is diffuse or directional should influence the method adopted for its measurement, as well as the calibration factor adopted for the instruments used.

The terrestrial radiation fluxes from the separate hemispheres of both the sky L_i and the earth's surface L_o are very substantially diffuse but not necessarily isotropic, although for the purposes of a general comparison of the relative magnitudes of the short- and long-wave fluxes, an assumption of approximate isotropy for the latter is acceptable. Again, for the purposes of comparison, typical terrestrial radiators might be supposed to have approximate temperatures of about 300 K.

Using Stefan's law, the ratio of the radiant intensities of the sun and the Earth is

$$\frac{I_{Earth}}{I_{Sun}} = \frac{I_{(T=300)}}{I_{(T=6000)}} = \left(\frac{300}{6000}\right)^4 = 6.67 \times 10^{-6} \tag{20.3}$$

However, when irradiances or flux densities are considered at a point on the earth, it must be considered that the solar intensity extends only over a solid angle of about 0.68×10^{-4} st, while the terrestrial intensity emanates from a solid angle of 2π so that if the former is considered to be normally incident then

$$F_{Earth} = \pi I_{Earth} \quad \text{and} \quad F_{Sun} = 0.68 \times 10^{-4} I_{Sun}$$

and consequently

$$\frac{F_{Earth}}{F_{Sun}} = \left(\frac{300}{6000}\right)^4 \frac{\pi}{0.68 \times 10^{-4}} = 0.29 \tag{20.4}$$

This ratio of about 0.29 is a useful guide to the minimum value to be expected, since solar attitudes below the zenith will result in lower values of F_{Sun} and hence higher values for the value of this ratio in equation (20.4). The solar and terrestrial irradiances are shown graphically in Figs 20.3, 20.4 and 20.5,

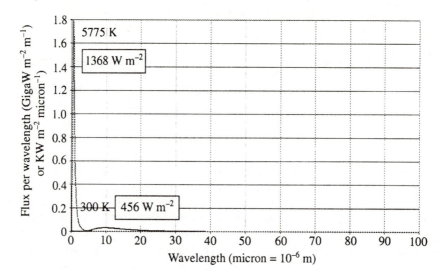

FIGURE 20.3 Solar and terrestrial radiation spectra compared using linear axes for both spectral flux and wavelength.

where the same information has been presented on respectively linear, log–linear and log–log sets of axes. It should be noted that the solar flux has been simulated by chosing an emission temperature of 5775 K rather than 6000 K since the total flux for the former more closely matches the actually measured value of the solar constant, i.e. the value of F_{Sun} at the 'top' of the atmosphere. The integrated total flux radiative density has been shown for both wavelength groups in all three figures.

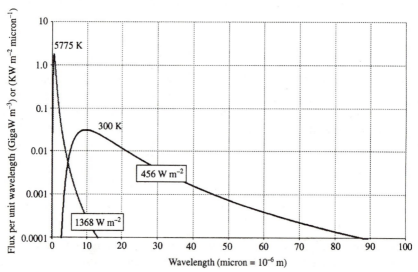

FIGURE 20.4 Solar and terrestrial radiation spectra compared using log spectral flux and linear wavelength axes.

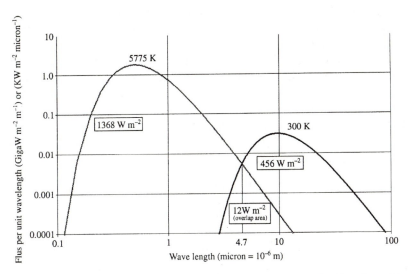

FIGURE 20.5 Solar and terrestrial radiation spectra compared using logarithmic scales for both flux and wavelength.

20.4 THE WAVELENGTH SEPARATION OF SOLAR AND TERRESTRIAL RADIATION

The separation of the solar and terrestrial wave-bands is apparent in all the figures 20.3, 20.4 and 20.5 but Fig. 20.4 allows the values of the coordinates at the point where the solar and terrestrial irradiances are equal to be read with some accuracy. In fact at a wavelength of 4.7 μm the solar irradiance is seen to be about 0.35% of its value at the maximum which occurs at a wavelength of about 5 μm. The data for all of these spectral graphs have been derived from spreadsheet computations based on the Planck spectrum equation (19.63). When this type of comparison is made for the terrestrial long-wave region, the result is not so favourable, in that at the same wavelength of 4.7 μm the spectral terrestrial irradiance (for 300 K) is about 6% of the value at the maximum. Perhaps more reassuring is to note that when the fluxes are computed, the total overlapping area accounts for only 12 W m^{-2}. Indeed, if 4.7 μm is chosen as the dividing wavelength, then 0.6% of the solar and 0.9% of the terrestrial components respectively are cut off. Were it possible to devise a sharp cut-off for both solar and terrestrial radiation sensors, at a wavelength of 4.7 μm, then it is clear that only a modest error would result.

It is clear that while there is a substantial degree of separation between the two flux spectra, the overlap should not be automatically ignored when long-wave, rather than all-wave radiometers are being used. In fact, the 'cut-off' wavelengths for the filters built into the domes of some modern pyrgeometers need to be considered before computing the total all-wave radiation for a given direction.

While spreadsheet-generated data have conveniently provided the data revealed in the first paragraph of this section, the Planck equation can readily be used in a simple analytical solution to find the value of λ at which the spectral solar and terrestrial fluxes are equal. As an example, this has been done for emitter temperatures of 5775 and 300 K, not forgetting the magnitudes of the effective 'apertures' of the two sources, which are 0.68×10^{-4} and π respectively, the ratio of which provides a 'normalizing' factor to enable a comparison of radiative flux density for terrestrial and solar sources at 300 and 5775 K respectively

$$\frac{8\pi hc}{\lambda^5} \frac{1}{\exp(hc/300\lambda k) - 1} = \frac{\pi}{0.68 \times 10^{-4}} \frac{8\pi hc}{\lambda^5} \frac{1}{\exp(hc/5775\lambda k)} \tag{20.5}$$

On substituting the by now known values of $c = 2.998 \times 10^8 \, \mathrm{m\,s^{-1}}$, $k = 1.38 \times 10^{-23} \, \mathrm{J\,K^{-1}}$ and $h = 6.625 \times 10^{-34} \, \mathrm{J\,s}$ into the above equation, it can be shown that $\lambda \approx 4.7 \times 10^{-6} \, \mathrm{m}$; that is, at a wavelength of $4.7 \, \mu\mathrm{m}$ the solar and terrestrial spectral fluxes are equal. It is clear that the graphical means of solution offered by means of a spreadsheet is far more convenient.

20.5 THE PLANETARY TEMPERATURE

Although the Earth experiences minor variations in its long-term temperature, as evidenced by both historical and palaeoclimatological records, there does exist a relatively stable thermal equilibrium. In order for this to have been maintained, the earth and its atmosphere must have attained some equilibrium temperature as a composite system radiating out to space, to balance the incoming radiation from the sun.

An elementary method of describing this process is achieved by considering the earth, when viewed from space, to have an overall mean albedo of β_E which as a further simplification is regarded as being independent of the angle of incidence of the solar irradiance, that is independent of the terrestrial latitude. Figure 20.6 illustrates this model of terrestrial irradiance F_{Sun} by the Sun. The total radiant power from the sun (rather than the radiant flux density) absorbed by the earth, E_{sun}, is found by annular integration

$$E_{\mathrm{Sun}} = 2\pi r^2 (1 - \beta_E) F_{\mathrm{Sun}} \int_0^{\pi/2} \sin\theta \cos\theta \, d\theta = \pi r^2 (1 - \beta_E) F_{\mathrm{Sun}} \tag{20.6}$$

where F_{sun} is the 'solar constant' expressed as a radiant flux density at the 'top' of the earth's atmosphere.

Equation (20.4) implies that in this simplified model, the Earth absorbs radiation from the sun as if the former were a plane circular disc of radius r and uniform albedo β_E.

On the other hand, this planet of radius r must be regarded as continuously radiating out to space from its entire effective radiating surface. The altitudes of the latter will be distributed throughout the atmosphere, but because the atmospheric thickness is small compared with the radius of the Earth, this area is $4\pi r^2$, at a total rate given by

$$E_{\mathrm{Earth}} = 4\pi r^2 \sigma T_{\mathrm{Earth}}^4$$

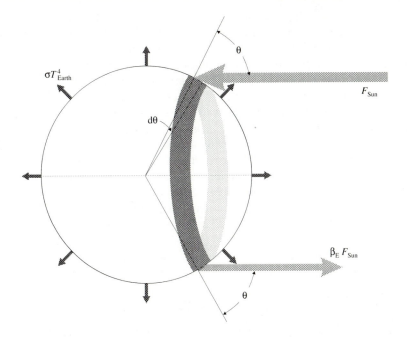

FIGURE 20.6 Simple model of the irradiance of the Earth by the Sun and the loss of radiation to space by the Earth.

where σ is Stefan's constant and T_{Earth} is the effective radiating temperature of the earth, a quantity which is known as the Earth's planetary temperature.

However, for the Earth as a whole to be in a state of radiative equilibrium, the rate at which energy is gained from the Sun must match the mean rate at which it is lost to space, that is $E_{\text{Sun}} = E_{\text{Earth}}$, and so from equations (20.4) and (20.5) it can be seen that

$$\pi r^2 (1 - \beta_{\text{E}}) F_{\text{Sun}} = 4\pi r^2 \sigma T_{\text{Earth}}^4 \tag{20.7}$$

Therefore

$$T_{\text{Earth}} = \left(\frac{(1 - \beta_{\text{E}}) F_{\text{Sun}}}{4\sigma} \right)^{1/4} \tag{20.8}$$

On substituting into equation (20.6) the observed value of the 'solar constant' of $F_{\text{Sun}} = 1370 \, \text{W m}^{-2}$ and the known value of Stefan's constant $\sigma = 5.67 \times 10^{-8} \, \text{W m}^{-2} \, \text{K}^{-4}$ as well as the estimated mean value of the terrestrial albedo $\beta = 0.34$, the Earth's planetary temperature is found to be 251.3 K which is about $-22°C$.

The planetary temperature can be regarded as an effective mean or equilibrium temperature which could be found to characterize any significant body in space as its effective overall temperature when determined by a relatively remote infrared or long-wave radiometer integrating the radiational characteristics over the entire surface. In the case of the Earth this could be envisaged as being done by an orbiting radiometer-equipped satellite. The atmospheric or terrestrial level of the effective radiating surface is of course not identified by a single, simple radiometric measurement, just as the Earth's albedo is determined in part by cloud and also by the Earth's surface itself where the sky is clear; that is, the effective albedo is distributed over a range of terrestrial altitudes.

It is interesting to compare the Earth's planetary temperature with the mean temperature of the earth's surface itself, which may vary from −40°C in polar ragions to 40°C in tropical latitudes. The overall, areally weighted mean is usually accepted as being 14.3°C which is much higher than the planetary temperature of −22°C. On the other hand, the latter is still much higher than the mean temperature of about −60°C at the tropopause. These facts suggest that the terrestrial radiation which is lost to space originates in part at the Earth's surface, but is contributed to by gaseous emitters right through the atmosphere to the highest levels. However, since it is known that only polyatomic gases, which are dominated in the lower atmosphere or troposphere by water vapour, are infrared or long-wave radiators, early theoretical models of terrestrial atmospheric radiation considered the tropopause as being the upper limit for contributing effective radiating surfaces.

20.6 SIMPLE MODELS OF THE GREENHOUSE EFFECT

Because part of the radiant flux lost to space is emitted at higher levels of the troposphere, this suggests a mechanism whereby the Earth's surface is protected from excessive radiative heat loss. The general nature of this mechanism can be illustrated by a simple example, in which the atmospheric gases participating in long-wave radiation are considered to be located in a compact layer at some height above the Earth's surface. This arrangement is shown in Fig. 20.7 in which this layer is shown to be transparent to the short-wave solar radiation of flux density F_{Sun} as is substantially the case in the absence of cloud. If the Earth's surface temperature is T_0 and that of the compact layer of long-wave radiating gases above it is T_1 then since all the latter radiates but, because of its short-wave transparency, relies on radiation from the earth's surface for all the radiant energy it receives, radiative equilibrium demands that $2\sigma T_1^4 = \sigma T_0^4$ so that

$$T_1 = \left(\tfrac{1}{2}\right)^{1/4} T_0 = 0.84 T_0 \qquad (20.9)$$

If this type of argument is extended to consider four compact layers of radiatively participating gases in the atmosphere, as illustrated in Fig. 20.8, then the following set of simple simultaneous equations can be solved

$$2T_4^4 = T_3^4$$

$$2T_3^4 = T_4^4 + T_2^4$$

$$2T_2^4 = T_3^4 + T_1^4$$

$$2T_1^4 = T_2^4 + T_0^4$$

Therefore

$$3T_4^4 = T_2^4 \quad 4T_4^4 = T_1^4 \quad \text{and} \quad 5T_4^4 = T_0^4$$

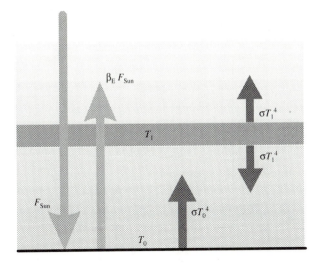

FIGURE 20.7 Model of a single-layer 'greenhouse'.

Inductively it then follows that for n layers, the temperature of the outermost or nth layer would be given by

$$T_n = \left(\frac{1}{n+1}\right)^{1/4} T_0 \tag{20.10}$$

Equation (20.10) implies that the rate of loss of energy to space would then be reduced by a factor of $1/(n+1)$ to the rate which would prevail in the absence of the atmospheric layers.

Measurements reveal that an atmospheric layer containing 0.3 mm of precipitable water has an effective emissivity of $\varepsilon = 0.2$ so that in this simple 'greenhouse' model, there would be a large number of relatively thin atmospheric layers in the lower troposphere, where the water vapour content tends to be high, and only a few, thick layers in the dry upper troposphere. The outermost of n atmospheric layers, each having 0.3 mm of precipitable water, would thus be radiating with a flux density and a rate given by

$$\varepsilon \sigma T_0^4 = \sigma \left(\frac{\varepsilon}{n+1}\right) T_0^4 \tag{20.11}$$

In this simple model, n needs to be found from an estimate of the total precipitable water in the atmosphere. On assuming an order of magnitude value of about 10 mm for the precipitable water in an atmospheric column, this implies a value of about 30 for n and consequently a ludicrously small value for the terrestrial radiant flux out to space. This false result suffices to show that the overly simple model of the atmosphere as a layered sequence of 'grey' body radiators is inappropriate, necessitating a more careful physical examination of the radiation processes.

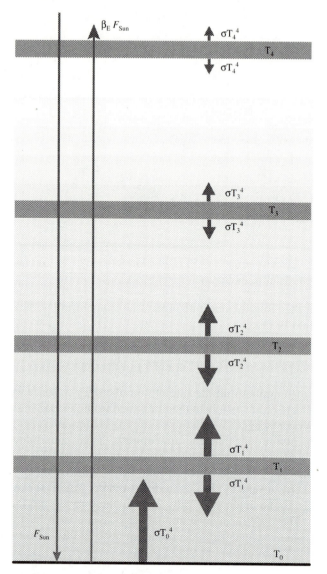

FIGURE 20.8 Model of a four-layer 'greenhouse'.

20.7 SIMPSON'S THEORY OF ATMOSPHERIC
RADIATION TRANSFER

Following the clear failure of the 'grey' radiational model of the atmosphere, in 1927 the English meteorologist George Simpson recognized the importance of considering the spectral dependence of atmospheric radiative properties. He was able to develop a relatively simple means for computing the radiation exchange between the surface of the Earth, its atmosphere and space.

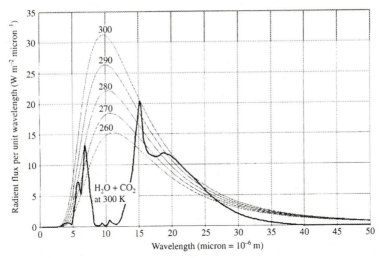

FIGURE 20.9 Planck spectra for temperatures between 270 and 310 K and the combined water vapour and carbon dioxide spectrum at 310 K.

Inspection of the infrared spectrum for a black body temperature of 300 K, for example in Fig. 20.4, indicates that practically all of the radiant energy lies between wavelengths of 3 and 50 μm for temperatures encountered in the atmosphere. Figure 20.9 shows the approximate form of the spectrum of radiation from an atmosphere of 'average' composition. Because the absorption bands of H_2O and CO_2 are substantially responsible for this radiation the great

Table 20.1 Slab emissivities (%) as a function of path length for water vapour and carbon dioxide at 20°C and standard sea-level pressure (after Elsasser and Culbertson, 1960)

Path length (mm)	H_2O	CO_2
1	35.92	3.01
2	40.18	4.21
5	46.71	6.15
10	52.81	7.70
20	60.26	9.26
50	72.39	11.30
100	81.38	12.80
200	88.04	14.27
500	91.44	16.19
1 000	–	17.62
2 000	–	19.01
5 000	–	20.83
10 000	–	22.17
20 000	–	23.48

range of concentrations of water vapour encountered in the atmosphere will cause the precise shape of the spectrum shown in Fig. 20.9 to vary. However, for a path length corresponding to the full 'thickness' of the atmosphere, the data in Tables 20.1 and 20.2 suggest that these variations are not large. In using the information from these tables, it is useful to imagine the gases in a 'real' vertical column of the atmosphere arranged as a column of uniform density, which would in fact be about 8000 m long at standard temperature and pressure. If CO_2 were present at a concentration of 0.03%, then its total radiative path length at STP would be about 2400 mm. What is important to note from Table 20.1 is that the doubling of this path length, that is doubling the atmospheric concentration, would only increase the CO_2 contribution to the radiant flux by about 5%. This contribution is in any case usually far smaller than that of water vapour, atmospheric path lengths for which in Table 20.1 typically range from 50 to 500 mm. This means that the total emissivity contributed by CO_2 is usually less than 20% of that for H_2O, where typical humidity ranges result in over 20% emissivity variations, with a doubling of CO_2 concentration producing an emissivity change of 5% of a 20% overall contribution, that is only 4%, which in turn is only 20% of total emissivity variations associated with normal water ranges of water vapour concentration in the atmosphere. The further water-vapour-induced changes in the atmospheric radiation budget owing to albedo responses to cloud formation mask the significance of CO_2 contributions still further.

Simpson was led to regard the atmospheric infrared spectrum in a manner summarized in Table 20.3. This table also incorporates data read from

Table 20.2 Slab emissivities (%) as a function of temperature for a constant path length of water vapour and carbon dioxide at standard sea-level pressure (after Elsasser and Culbertson, 1960)

Temperature	H_2O (20 mm path length)	CO_2 (2 m path length)
−80	70.00	15.91
−70	67.10	16.71
−60	64.85	17.38
−50	63.14	17.94
−40	61.87	18.38
−30	60.97	18.71
−20	60.38	18.93
−10	60.05	19.07
0	59.94	19.12
10	60.02	19.09
20	60.26	19.01
30	60.63	18.87
40	61.12	16.68

Table 20.3 The radiation losses to space from the earth–atmosphere system

Wavelength range (μm)	Terrestrial origin of radiation	Radiation loss to space per unit area (W m^{-2})
0–7	Stratosphere	1
7–8.5	Mean of surface and stratosphere	18
8.5–11	Earth's surface	63
11–14	Mean of surface and stratosphere	45
14–100	Stratosphere	84
0–100	**Composite terrestrial system**	**211 (composite terrestrial flux)**

Simpson's spectral model shown in Fig. 20.10 which he devised by considering that any radiation originating at the earth's surface in the region between 8.5 and 11 μm will escape to outer space because of the transparency in this region shown in Fig. 20.10. For these wavelengths, the energy lost to space is thus proportional to the area under the spectrum for a black-body at the earth's surface temperature. For the regions below 7 μm and above 14 μm the theory of the grey body atmosphere is applied and the last atmospheric layer with 0.3 mm of precipitable water determines the radiation into space. For the small inter-mediate-wavelength regions, i.e. 7 to 8.5 μm and 11 to 14 μm, Simpson used graphical interpolation to indicate the total radiated energy. Spreadsheet techniques now allow this to be accomplished very simply, in particular the calculation of the areas which represent radiated powers, for which results are shown in Table 20.3.

This summary of the effective atmospheric (i.e. long-wave) radiation out to space can now be used in a comparison with the power of the short-wave

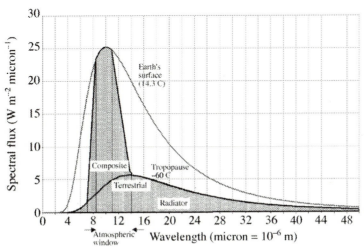

FIGURE 20.10 Simpson's model of the atmospheric 'window'.

radiation gained by the whole Earth from the Sun which was calculated in the discussion of the planetary temperature.

The essential details are noted in Table 20.4 which reveals the remarkable accuracy achieved by Simpson in describing the mechanism of the terrestrial radiation balance. In the calculation of the solar irradiance of the Earth's surface, the solar flux (solar constant) is taken to be $1370 \, \text{W} \, \text{m}^{-2}$ and the value used for the mean terrestrial albedo is 34%. While more detailed knowledge of the spectra of the polyatomic atmospheric gases is now used in more elaborate radiation models of the atmosphere, the Simpson theory clearly establishes the basic nature of the fundamental mechanisms.

21

CLIMATE CHANGE

21.1 INTRODUCTION

Climate has been defined by meteorologists as average weather. The Oxford dictionary defines it as 'a place's weather characteristics'. These are rather simple definitions. Climate might also be defined as the ensemble of all relevant statistics for all meteorological observations averaged over the time period and the location, area or region to which the term climate refers. If we consider temperature we might regard the annual mean temperature of a place or geographical region calculated for a period of a decade or more to be its climate in so far as temperature is concerned. But each month of the year would also have its climate. In some places winters might be very cold and summers very hot, yet the mean annual temperature might be the same as if every month had the same mean temperature. In the latter case the range and the variance of the temperature would be important factors, in addition to the mean. Similar considerations would apply to other meteorological parameters such as rainfall, cloudiness, humidity, wind strength, etc.

21.2 DEFINITIONS

A somewhat formal definition of climate used by the WMO is

> Climate is the synthesis of weather over the whole of a period essentially long enough to establish its statistical ensemble properties (mean values, variances, probabilities of extreme events, etc.) and is largely independent of any instantaneous state.

Weather changes from day to day. The average weather for a given month one year may be different from that in the same month in another year. The average weather for a given year may be different from the average weather in another year. Has the climate changed? Or has the climate varied? The word changed in this context implies a permanent change. The word varied implies a temporary change from one condition to another and then back again. The expression 'climate change' has been used a great deal during the past 10 years and has become identified with an absolute truth. What is meant by 'climate change'?

This perplexing question can perhaps best be answered by referring to a footnote contained in the Australian delegation's report of the November 1995 meeting of the Working Group (Science) of the WMO/UNEP Intergovernmental Panel for Climate Change (IPCC), and subsequently included in the WMO/UNEP publication *Climate Change 1995*. It states:

> Climate Change in IPCC Working Group I usage refers to any change in climate over time whether due to natural variability or as a result of human activity. This differs from the usage in the Framework Convention on Climate Change where Climate Change refers to a change in climate which is attributed directly or indirectly to human activity that alters the composition of the global atmosphere and which is in addition to natural climate variability observed over comparable time periods.

21.3 GLOBAL WARMING

The basis of any conjectured human-induced climate change within the near future must first lie in the concept that an enhanced greenhouse effect induced by human activities must warm the planet to an extent which may be detected, and generally agreed to lie outside the range of normal climate variability. This concept has become known by the term 'global warming'. In any ensuing debate about human-induced climate change the factual existence of global warming should be the main issue in the first instance. Any redistribution of the three-dimensional pressure field, that is any changes in the general circulation of the atmosphere and/or in the frequency of severe storms or weather events which might occur as a consequence of an overall warming, is a second-order effect and one of great complexity. This is a problem which has to be addressed by high-resolution numerical models, the results from which must be carefully interpreted and cannot always be taken on their face value.

21.4 CLIMATE VARIABILITY

Climate variability may be regarded as the sum of natural variability and artificially induced variability. Natural climate variability may be interpreted as the spatial and temporal changes in observed measurements of atmospheric quantities such as temperature, precipitation, cloudiness and many others which occur as a consequence of the internal mechanisms of the ocean–atmosphere system. This variability is the observed result of the non-linear internal response to the external forcing of solar radiation, and perhaps volcanic eruptions. Although the latter forcing is not external to the planet it is external to the ocean–atmosphere fluid. Artificially induced climate variability may be regarded as changes in climate which result from human-induced processes such as the alteration of the constituents of the atmosphere due to the release of various chemical compounds such as are caused by carbon emissions, and other effluents from industrial and domestic appliances. The most important among such artificial perturbations to the ocean–atmosphere system has come to be known as the enhanced greenhouse effect.

The concept of climate change in the sense of a human-induced and permanent change in the statistics of climate has been in the public eye from time to time in the past, but it has only been during the last 10 years that it has commanded worldwide public and political attention, leading to international action addressed to the problem of reducing emissions of carbon dioxide and other greenhouse gases.

21.5 THE GREENHOUSE EFFECT

The analogy between a garden greenhouse and global warming can only be made in a very crude way. Although the garden greenhouse is transparent to short-wave solar radiation and traps the longer infrared radiation, it is also enclosed by walls and, more importantly, by a roof. Heat cannot escape through the roof by convection as in the real atmosphere. A greenhouse can therefore become much warmer than its external environment. In the atmosphere the greenhouse effect arises from the fact that certain gases are transparent to short-wave solar insolation but are opaque to long-wave terrestrial radiation. However, the gases do not constitute a solid roof such as is provided by the glass of the garden greenhouse, or by the closed glass windows and sun roof of a car. There can be no transport of heat by convection from the glass greenhouse or motor car to the external environment. Thus, although the use of the simile 'greenhouse' is quite apt and useful, it should not be interpreted too literally in the common garden sense.

The most important greenhouse gases in the atmosphere are water vapour and carbon dioxide in that order. Now the amount of water vapour in the atmosphere remains fairly constant. Evaporation is balanced by precipitation, processes which form part of the hydrological cycle. However, this is not the case for carbon dioxide which does not liquefy at temperatures encountered naturally on the earth. The content of carbon dioxide in the atmosphere has increased from about 280 parts per million by volume before 1800, to about 360 ppmv today. This is believed to be almost entirely due to emissions for which human beings are responsible in consequence of their needs for energy and consumer goods and services, including agricultural practices which alter the surface of the planet. A specific example is the reliance of inhabitants of the industrialized nations on the motor vehicle.

It has been said (IPCC, 1990) that the earth is about 33 K warmer than it would be if there was no greenhouse effect. This is obtained from the formula for the radiative equilibrium temperature

$$T^4 = \frac{(1-a)Q}{4\sigma} \tag{21.1}$$

where Q is the solar constant, about 1360 W m^{-2}, a is the albedo, the fraction of solar isolation reflected back to space by the earth, and σ is 5.67 $\times$ 10^{-8} W m^{-2} K^{-4}. Substituting these values and 0.30 for a in equation (21.1) we find that $T = 254.5\,\text{K}$ or about $-18°\text{C}$. We know that the mean observed temperature of the earth is about $288\,\text{K}$ or $15°\text{C}$. However, in making this calculation we have assumed an albedo of 0.30. Since the latter value is largely due to reflection from the tops of clouds, and clouds are composed of condensed water vapour, we must assume there is water vapour in the air. But we have also assumed there is no absorption by the water vapour which would have to be in the atmosphere if there were clouds. As pointed out by Professor R. S. Lindzen of MIT (1991), the calculation is not really valid. If you throw away the water vapour you also throw away the clouds. If we substitute zero for the albedo we find that $T = 278\,\text{K}$ or $5°\text{C}$, still $10°\text{C}$ colder than observed but certainly not everywhere icebound. However, the surface of a totally dry earth would still have an albedo. Supposing in a dry world one assumes an albedo of, say, 0.15, we then find that the mean temperature of the earth would be about $-6°\text{C}$; that is, about $21°\text{C}$ colder than it actually is.

Comparisons are sometimes made with Venus. The temperature there is observed to be about $760\,\text{K}$, in spite of an albedo of 0.77. On the other hand, the solar constant is about double that for the earth, the atmosphere is almost wholly carbon dioxide and the surface pressure is 90 times greater than on the earth. Thus, with a composition of carbon dioxide some 180 000 times more than on the earth it is clear that a strong greenhouse effect must exist, but it can hardly be a fair comparison with the earth, where the human-induced radiative forcing has been about $2.5\,\text{W}\,\text{m}^{-2}$ since the beginning of the industrial age in the middle of the nineteenth century.

21.6 THE OBSERVED GLOBAL TEMPERATURE RECORD

The physics of radiative forcing can tell us the theoretical amount by which the planet can be warmed. But that physics alone does not include any complex feedbacks. There are two methods of deducing information about a possible 'enhanced' greenhouse effect. The first is to examine and analyse the time series of past records of temperature data. The second is to build theoretical models incorporating most of the equations we have studied, and others. The models simulate the behaviour of the real atmosphere by means of computer calculations for successive time intervals. They start to run at some initial time for which actual or assumed numerical values are assumed valid. Both methods have inherent difficulties in divulging the real truth and nature of what happens in the real world.

There are several data banks of hemispheric and global series of mean temperatures. All the series are expressed as anomalies from some mean reference value. They also indicate the same general result that the mean global surface temperature has increased from about 0.3 to $0.6°\text{C}$ during the past 100 years or so, but the increase has not been continuous. This is not a linear relation with time, although the trend computed as a linear regression may be shown as a straight line.

Figure 21.1 shows the global temperature record for the period 1856–1996. The values represent anomalies of temperature from the mean for the period

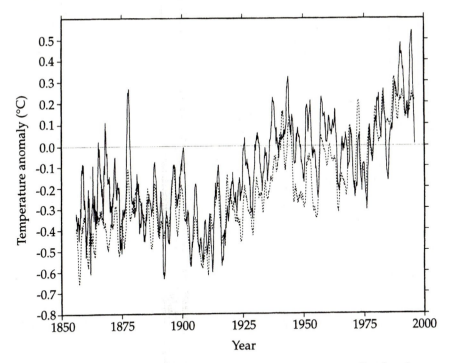

FIGURE 21.1 Twelve month running average temperature anomalies for the two hemispheres (1856–1996): southern hemisphere, dashed line; northern hemisphere, solid line.

1961–90, computed from the internationally recognized Parker/Jones historical temperature record. The curve is a 12 month running average. It shows an increase of global 12 month mean surface air temperature between 1880 and 1940, a cooling from 1940 to about 1970 and then another sharp warming up to 1996 (although there has been a cooling of about 0.2°C during 1996 in the running average).

21.7 RANDOM WALKS

The shape of temperature curves such as shown in Fig. 21.1 has often led to the comment that 7 or 8 years, or whatever it may be at the time, out of the past 10 years have been the warmest of the whole historical record. Superficially, this fact seems to present a strong argument of some trend that is being forced continually, but mathematically the argument is unconvincing.

Pure random walks may give the impression that there is a forcing trend when actually the series is composed by pure chance, as during the tossing of a fair coin. Figures 21.2 show examples of five random walks which were generated by a computer for first-order differences possessing a normal distribution of mean 0 and standard deviation of the same magnitude as possessed by the real-world series. It is noted that the excursions within the same timeframe generally

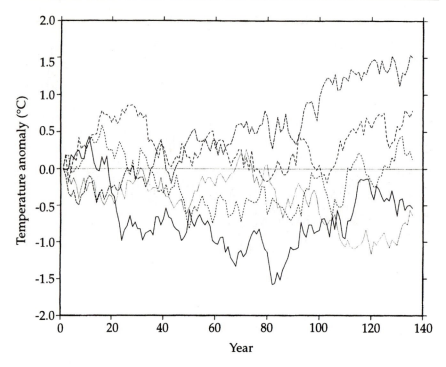

FIGURE 21.2 An ensemble of random walks generated with the same standard deviation statistic as the historical temperature series.

exceed those shown by the global temperature series in Fig. 21.1. The real-world temperature trend appears to contain a natural variability that possesses some random walk characteristics, but these are not so well developed as in the theoretical case. Although the variances of the first-order differences used to generate the walks and those of the real-world series are the same, the variances of the random walks themselves are several times larger than the variances of the real-world series.

The curve illustrated by δ/λ the top curve in Fig. 21.2 is of special interest. If observed in a series of real observations the warming trend would almost certainly be interpreted as having been forced by some external factor. In order for the global temperature series to be transformed into a true random walk the interannual changes should be normally distributed and the energy of each change must be conserved for all time. That is an unrealistic assumption to make. However, for short time scales, such as the length of the period of historical temperature records, some randomness is inherent, and care must be exercised in drawing conclusions. For these reasons the pre-1996 official point of view put forward by the WMO is

it is still not possible to attribute with high confidence all, or even a large part of, the observed global warming to the enhanced greenhouse effect. On the other hand, it is not possible to refute the claim that greenhouse-gas-induced climate change has contributed substantially to the observed warming.

21.8 THE DEBATE

During 1986–90 the greenhouse effect debate raged furiously. Various greenhouse action societies were formed. Warnings of impending catastrophe were promulgated in the press and electronic media. On the other side there were equally mistaken views which maintained that the physics had been interpreted incorrectly, and that little or no effect would occur. In those years there seemed to be little coordination or authoritative source of information.

The publication of the IPCC (1990) report entitled *Climate Change 1990: The IPCC Scientific Assessment* was a milestone in the debate. This report of some 350 pages consisted of 11 chapters and was compiled by more than 30 authors, while nearly 300 scientists contributed to the final text. It was sponsored jointly by the WMO and the United Nations Environment Programme (UNEP). As such it was regarded by many (both governmental bodies and non-governmental groups and by the public at large) as representing the final official and authoritative point of view on climate change and global warming. Its main conclusions were that under a 'business as usual scenario' with no controls over CO_2 emissions the rate of increase of temperature would be about 0.3°C per decade, leading to an increase of 3°C before the end of the next century. This amount of warming would lead to an average increase of mean sea level of about 6 cm per decade over the next century, mainly due to thermal expansion of the oceans and melting of some land ice. This predicted rise would reach about 65 cm by the end of the next century. The IPCC (1990) report was updated in 1992 with a further report on climate change (IPCC, 1992). The latter came to the same general conclusions as the earlier report but added to and amended the earlier version in minor detail on the basis of later research and opinion. In the past few years, several model experiments have shown that even if greenhouse gas emissions were to be stabilized, sea level will continue to rise for decades if not for centuries (IPCC, 1996). Finally, at the time of writing the new 1996 update has been published. This recent addition to the debate will be discussed in the final section of this chapter.

The first IPCC report did not escape criticism, in spite of the eminent stature of its authors and contributors (Lindzen, 1991).

21.9 THE MSU DATA

MSU stands for microwave sounding units. The units are carried by satellites and they scan the atmosphere as they travel in their orbits around the planet. This passive microwave radiometry from satellites provides a more precise record of atmospheric temperatures than do the surface historical records such as shown in Fig. 21.1. The first major announcement of temperature trends computed from the MSU data was contained in a paper by Spencer and Christy which was published in March 1990. It made something of an impact and was included in television news broadcasts. The observational period was from 1979 to 1988, since extended to 1996. The 1990 paper reported that

> analysis of the first 10 years (1979–1988) of satellite measurements of lower atmospheric temperature change reveals a monthly precision of 0.01°C, a large temperature variability on time scales from weeks to several years, but no obvious trend for the 10 year period. The years 1980, 1983, 1987 and 1988 were the warmest for the period while 1984, 1985 and 1986 were coolest.

Figure 21.3 shows a comparison of the satellite-monitored temperature anomalies for the two hemispheres for the period January 1979 to March 1997. The sharp cooling which occurred in 1991–2 is due to volcanic dust ejected from the eruption of Mt Pinatubo in the Philippines in June 1991. Figure 21.4 shows the 12 month running averages of the curves shown in Fig. 21.2. The plot is much smoother than the plot of the actual monthly observations. Figure 21.5 compares the time series of the surface data and the satellite data over the same period for the globe. It is noted that although there is no warming trend in the satellite observations, the coherence between each set of data is very high.

Table 21.1 shows the extremes for the monthly MSU values, extended to December 1995. The warmest month for the globe and for the northern hemisphere occurred in December 1987. It came at the end of a warm phase of ENSO, a phenomenon which will be described in the following section. The

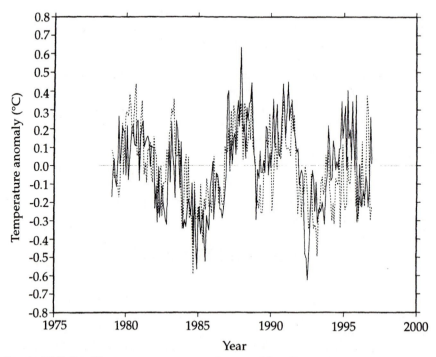

FIGURE 21.3 Satellite temperature anomaly record for the northern (solid line) and southern (dashed line) hemispheres (January 1979 to March 1997).

Table 21.1 Maximum and minimum temperatures for the globe and hemispheres for the period 1979–95 inclusive for the monthly series of MSU data (anomalies from the mean for the period)

	Max	**Date**	**Min**	**Date**
Globe	0.472	Dec. 87	−0.458	Sept. 92
NH	0.624	Dec. 87	−0.652	July 92
SH	0.417	Sept. 80	−0.537	Mar. 93

coldest days occurred in different years. The coldest month of all was in July 1992 for the northern hemisphere, a little more than a year after the volcanic eruption of Mt Pinatubo. The southern hemisphere reached a minimum in March 1993.

Table 21.2 shows changes of mean temperature for the globe and the hemispheres from one month to the next.

21.10 THE ENSO PHENOMENON

ENSO stands for El Nino Southern Oscillation. This phenomenon is regarded by meteorologists and oceanographers as the most significant air–sea interaction event with respect to its influence on climate on a scale of 3 to 6 years. El Nino is

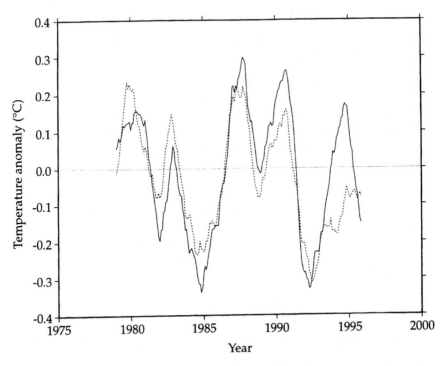

FIGURE 21.4 Twelve month running satellite-monitored temperature for the two hemispheres (January 1979 to March 1997): southern hemisphere, dashed line; northern hemisphere, solid line.

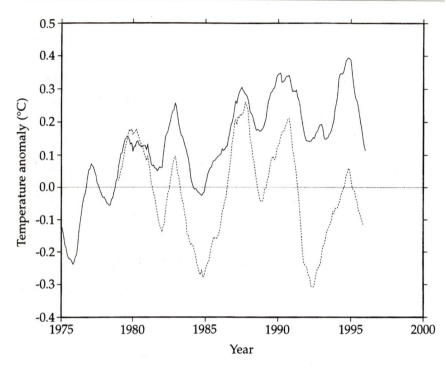

FIGURE 21.5 Twelve month running average temperature anomalies for the global land marine temperature (solid line) and the global satellite temperature (dashed line).

the name of a warm ocean current which replaces the normally cold water off the coast of Peru. The Southern Oscillation accompanies the oceanic component and is measured by an index, (the Troup index, SOI), which is the normalized difference in m.s.l. pressure anomaly between Tahiti and Darwin. Thus if the m.s.l pressure difference anomaly Tahiti–Darwin is positive ENSO is in its more normal mode; if the difference anomaly is negative, ENSO is in its El Nino mode. The effect on the climate is very dependent on the magnitude of the difference. An ENSO or El Nino event refers to a relatively large negative value of the SOI. When this happens surface water temperatures of the eastern and central tropical Pacific Ocean become abnormally high. Figure 21.6 (Figure 10.6 on page 300 of Neil Wells' text *Atmosphere and Ocean*) brings this factor out well. Water temperatures in subtropical latitudes in the southern hemisphere become less

Table 21.2 Greatest rises and falls of temperature from one month to the next for the 1979–95 series of daily MSU data

	Max rise	Date	Max fall	Date
Globe	0.395	Sept. 84	−0.270	Nov. 95
NH	0.401	June 83	−0.710	Nov. 95
SH	0.514	Sept. 84	−0.425	Sept. 94

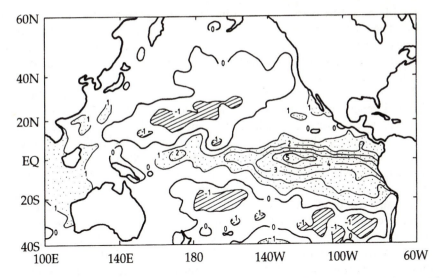

FIGURE 21.6 Sea surface temperature (°C) in the tropical Pacific Ocean during an El Nino event (December 1982).

than average and this may result in drought or long dry periods in eastern Australia.

21.11 NUMERICAL MODELLING OF THE CLIMATE

Numerical general circulation models (GCMs) were first developed in the 1950s, mainly in the USA (see Phillips, 1956). However, it was not until the 1960s when full-scale GCMs became major research projects within special geophysical fluid dynamics laboratories. Since that time the science of modelling has become what might be termed a growth industry.

A GCM attempts to develop the existing climate, solving sets of partial differential equations which incorporate the physical mechanisms and processes that are in continual operation in the atmosphere–ocean system. We have derived many of the basic mathematical equations which have already been used in this text. Others are too advanced to be included here. It is first necessary to set the initial conditions. One set of initial conditions might be to consider the atmosphere to be stationary and the pressure gradient fields zero at all elevations. The sun is then switched on and the atmospheric temperature, pressure and wind distribution slowly develop as the model is run with some selected time interval for the integration. Alternatively, one might select some climatic mean values for the initial conditions. Such models will eventually reach a state of equilibrium, which resembles the observed climatic distribution of the variables computed. This model may then be used as a control model for comparison with others which embody different initial conditions or contain parameters which change throughout the numerical integration. During the past 30 years there have been enormous developmemts in these GCMs, which now incorporate a coupling mechanism with the oceans.

GCMs have been very successful in reproducing the general climate of the globe. They have not been so successful in predicting, in advance, changes in the mean state, as, for example, from one season to the next or from one year to the next. The success in simulating the broad climate may to some extent be due to tuning the model to reproduce what is observed.

The early GCMs were mainly concerned with attempting to reproduce the observed general circulation, or climate. Attempts were also made to predict seasonal or monthly anomalies, but such predictions were never very successful. During the past 10 years attention has been directed towards the formulation of models which will predict the extent of global warming expected to occur as a direct result of the enhanced greenhouse effect. There are two types of models which may be used for this purpose: the equilibrium model and the transient model. Both types must assume a selected rate of increase of CO_2. The target is to predict the warming which would occur if some existing or preindustrial CO_2 concentration doubled. It is therefore necessary to surmise the rate of increase of CO_2. This is done by supposing different scenarios, based on assumptions of more efficient energy use in the future. The worst scenario is usually called the 'business as usual scenario', the meaning of which is self-evident.

21.11.1 The equilibrium model

The equilibrium model attempts to predict the warming which would occur if the concentration of CO_2 were doubled instantaneously. The model is therefore similar to a normal GCM with the imposition of the condition of a doubling of CO_2. Other initial conditions remain the same. The mathematical expressions for the radiative forcing embodied in the model compute a new temperature and the model incorporates the new information in its output. Eventually the model settles down into a state of equilibrium. One of the factors which complicates the difficulty of identifying a result believed to be caused by a real physical forcing, such as that induced by greenhouse gases, is that there is natural or random variability in the changes which occur with time. This natural variability is inherent in both the real world and in the GCM control models which try to simulate the real world. Sometimes the natural variability is called noise. It is not possible to predict the noise accurately as a function of time although GCMs do produce noise in their results. Some of the criteria against which GCMs must be tested concern comparisons between the statistics of the natural variability and the noise generated by the model simulation. Of course, models are tuned to ensure that there are no wide differences between these statistics, as, for example, the standard deviation of the modelled time series of temperature anomalies.

Thus, when the equilibrium model has settled down into the new condition of double CO_2 a mean temperature over some standard period is computed. The original temperature may then be subtracted from the new double CO_2 equilibrium temperature to give the global warming for whatever period has been assumed as the doubling CO_2 period.

21.11.2 The transient model

The equilibrium model is more economic to run as it does not require as much computer time. In fact, it is only necessary to run two GCMs side by side and compare their equilibrium states. However, the atmosphere does not work like that. Changes are not discontinuous but continuous. The transient model attempts to overcome this problem and tries to reproduce more closely what happens in the real world. Thus the CO_2 concentration is increased gradually. The resulting integrations are therefore much longer and so is the computer time required to integrate in small time steps some hundreds of years into the future. The latest results at the time of writing show a nearly linear trend of about 0.3°C per decade, depending on the assumed CO_2 doubling period. These results show warmings about 60% of the equilibrium model results. The difference between the results of the equilibrium and transient models is partly due to the inclusion of the deep ocean in transient models. The inertia of the oceans is considerable so that equilibrium is not reached when double CO_2 is reached.

The models attempt to predict other meteorological parameters in addition to temperature. Specifically, they attempt to predict these parameters as a function of the planetary geography. It might be thought, however, that the prime problem is first to establish the magnitude of hemispheric and global warming. Too much attention and policy directives towards exact geographical mapping of changes in climate parameters may not be rewarding at the present stage of knowledge, although a great deal of effort is being put into this approach.

21.12 THE GLOBAL WARMING DEBATE CONTINUES

In March 1996 the volume *Climate Change 1995* (IPCC, 1996) was distributed. This report continued the IPCC Scientific Assessment of Climate Change begun with the 1990 and 1992 reports. The essential difference between the 1995 and earlier assessments was the inclusion of a bottom-line statement which concluded that 'the balance of evidence suggests that there is a discernible influence on global climate'. This was the statement that the world had been waiting for. Governments needed such a statement to justify the action and pressure that had been taken worldwide to reduce CO_2 emissions.

The statement was based on the results of an investigation embodied in chapter 8 of the 1995 report, a chapter entitled 'Detection of Climate Change and Attribution of Causes'. The research consisted of an analysis of radiosonde observations as a function of latitude and height. Using the period 1963–1987 it was found that there was a pronounced warming trend in the southern hemisphere, which was not matched by a similar magnitude of trend in the northern hemisphere. The substance of chapter 8 was published in the scientific journal *Nature* in July 1996. At the time of writing responses to the published article had not been printed. The global debate continues.

21.13 CLIMATE PREDICTION

The goal of predicting the weather far into the future has always been an elusive dream. The accuracy of forecasts based on synoptic analysis of surface and upper air charts decays rapidly after about 24 hours. Forecasts computed from complex high-speed and high-resolution numerical models can now be extended from 5 to 10 days and retain useful information, although the degree of accuracy relies a great deal on the type of synoptic pattern prevailing. Large anticyclonic blocks are stable and persist for some time, whereas periods of intense baroclinic activity are dynamically unstable and lead to the problem of sensitivity to initial conditions and resulting chaos.

After World War II the idea of long-range forecasting received increasing attention. Indeed there had been enormous effort in this field leading up to the Normandy landings in June 1944. The general technique adopted was called the analogue method. This relies on selecting a chart from a long period of historical records of previous synoptic charts and selecting one that appeared most closely identical with the one for the present time, on which the forecast had to be based. It was assumed that the current chart would develop in a similar manner to the analogue chart chosen from some past date. The aim of the Normandy landing weather forecast exercise to produce a prediction for several days ahead was more closely related to the aims of the medium-range forecasts which have been produced for many years now by the European Centre for Medium-Range Weather Forecasting (ECMW) at Reading in England. Climate prediction is something different. It involves prediction of the mean state for a month, several months or even some years into the future.

In the 1950s the analogue method was used to try and predict the overall mean state, warm or cold, wet or dry, of the following month. In this procedure the analogue consisted of a chart of the mean pressure field selected from past records which most closely resembled the mean pressure field of the month immediately past. Thus, on or about the first day of each calendar month a forecast was presented for that month. The exercise was carried out for many years on an experimental and trial basis by the UK MO. Finally, it was decided to issue the monthly predictions officially as a service to the public. The practice lasted some 10 years or more, but it was eventually discontinued as statistical analyses suggested the results were no better than persistence or chance.

Today climate prediction is a problem that once again has come to the fore, largely driven by the need to make predictions of global warming caused by the 'enhanced' greenhouse effect. But although it may be possible to predict different climates which have different solar radiation inputs, different albedos or different CO_2 concentrations, it may well be impossible to predict changes in our climate under the assumption that the various forcings remain invariant.

Any progress towards even a partial solution to the problem rests on two possible techniques. The first is the harnessing of high-speed computers to the development of coupled ocean–atmosphere numerical models, the second is the use of various statistical techniques. To some extent both these roads depend on the high heat capacity of the oceans and the sluggish response of ocean circulations to forcings. The ENSO phenomenon discussed earlier is a very important link and may help to introduce occasional good results up to a few years ahead.

Prediction using the persistence of climatic features will always be better than pure chance because an even-odds choice means there is no persistence.

No improvement in climate prediction can be realized without an evaluation of results predicted in advance, not in arrears. New statistical techniques such as *cross-validation* and *bootstrapping* evade this problem to some extent but still implicitly assume that the next era will exhibit the previously observed statistical relationships. Nevertheless, as most statistical significance tests need a fairly large number of trials it is doubtful that any twentieth-century reader will see the realization of the dream.

21.14 PROBLEMS

1. Suppose the time constant (e-folding period) for k is 3.5 years, a value which may be attributed to the oceans, and ω is 26.4 months, a mean value for the quasi-biennial oscillation. Write a computer program and plot the lagging temperature, θ, of the oceans if the atmosphere is subjected to a sinusoidal QBO oscillation of amplitude 0.1°C. Assume the Newtonian expression $d(\theta)/dt = -k(T - \theta)$ is applicable. Try 1 month time intervals. Now assume that successive QBOs have varying periods (ω) and plot the results of temperature against time.

2. Suppose the atmospheric temperature increases by 0.3°C per decade as a result of greenhouse-induced warming. Assume a lag time constant of 3.5 years. How much do the oceans lag the atmospheric temperature as time becomes infinitely great?

BIBLIOGRAPHY

Abbott, P. F. and Tabony, R. C. 1985 The estimation of humidity parameters. *Meteorological Magazine*, **114**, 49–56.

Bennets, D. A., Grant, J. R. and McCallum, E. 1988 An introductory review of fronts, part 1: theory and observations. *The Meteorological Magazine*, **117**, 367–80.

Bluestein, H. B. 1992 *Synoptic-dynamic meteorology in midlatitudes*. Oxford: Oxford University Press.

BMTC 1988 *Satellite cloud imagery guide*. Bureau of Meteorology, Australia.

Browning, K. A. 1985 Conceptual models of precipitation systems. UKMO RRL Research Report no. 53.

Bureau of Meteorology 1993 Research papers II. Bureau of Meteorology, Northern Territory, Australia.

Cadet, D. and Overlez, H. 1976 Low level air circulation over the Arabian sea during the summer monsoon as deduced from satellite-tracked super-pressure balloons. Part 1. *Quarterly Journal of the Royal Meteorological Society*, **102**, 805–16.

Carlson, T. N. 1991 *Mid-latitude weather systems*. New York: McGraw-Hill.

Charney, J. G. 1948 On the scale of atmospheric motions. *Geos Physics Publication*, no. 17.

Climate Analysis Center (US) 1986 *Atlas of the tropical and subtropical circulation derived from National Meteorological Center operational analyses*, ed. Phillip A. Arkin *et al*. Silver Spring, MD: NOAA, National Weather Service, NOAA atlas no. 7.

Coulson, K. L. 1975 *Solar and terrestrial radiation*. New York: Academic Press.

Djuric, Dusan 1994 *Weather analysis*. Englewood Cliffs, NJ: Prentice Hall.

Dutton, J. A. 1976 *The ceaseless wind*. New York: McGraw-Hill.

Dvorak, V. F. 1975 Tropical cyclone intensity analysis and forecasting from satellite imagery. *Monthly Weather Review*, **103**, 420–30.

Dvorak, V. F. 1984 Tropical cyclone intensity analysis using satellite data. NOAA Technical Report NESDIS 11, 47 pp.

Elsasser, W. M. and Culbertson, M. F. 1960 Atmospheric radiation tables. *AMS Meteorological Monographs*, vol. 4, no. 23. Boston: AMS.

Emanuel, K. A. 1988 The maximum intensity of hurricanes. *Journal of Atmospheric Science*, **45**, 1143–55.

Ferriére, P. J. 1994 Weather hazards at Adelaide Airport. Met. Note 204, Bureau of Meteorology Melbourne, 35 pp.

Foley, G. R. and Hanstrum, B. N. 1994 The capture of tropical cyclones by cold fronts off the west coast of Australia. *Weather and Forecasting*, **9**, 577–92.

Fraedrich, K. and McBride, J. L. 1995 Large scale convective instability revisited. *Journal of Atmospheric Science*, **52**, 1914–23.

Frank, W. M. 1987 Tropical cyclone formation. Chapter 3 in *A global view of tropical cyclones*. Washington, DC: Office of Naval Research, 53–90.

Fredricksen, J. S. 1984 The onset of blocking and cyclogenesis, onset-of-blocking and mature anomalies. *Journal of Atmospheric Science*, **44**, 2562–74.

French, A. P. 1958 *Principles of modern physics*. New York: John Wiley.

Gray, W. M. 1968 Global view of the origin of tropical disturbances and storms. *Monthly Weather Reviews*, **96**, 669–770.

Haltiner, G. J. 1971 *Numerical weather prediction*. New York: John Wiley.

Hercus, E. O. 1950 *Elements of thermodynamics and statistical mechanics*. Melbourne: Melbourne University Press.

Holland, G. J. 1982 An analytic model of the wind and pressure profiles in hurricanes. *Monthly Weather Review*, **108**, 1212–18.

Holland, G. J. 1987 Mature structure and structure change. Chapter 2 in *A global view of tropical cyclones*. Washington, DC: Office of Naval Research, 13–52.

Holland, G. J. (ed.) 1993 *Global guide to tropical cyclone forecasting*, WMO/TD 560, TCP Report 31, Geneva: WMO/Tropical Cyclone Programme.

Holton, J. R. 1979 *An introduction to dynamic meteorology*, 2nd edn. London: Academic Press, 391 pp.

Holten, J. R. 1992 *An introduction to dynamic meteorology*, International Geophysics Series, vol. 48. San Diego: Academic Press.

Hoskins, B. J. 1975 The geostrophic momentum approximation and the semi-geostrophic equations. *Journal of Atmospheric Science*, **32**, 233–42.

Hoskins, B. J. and Bretherton, F. F. 1972 Atmospheric frontogenesis models: mathematical formulation and solution. *Journal of Atmospheric Science* **29**, 11–37.

Hoskins, B. J. and Pedder, M. A. 1980 The diagnosis of middle latitude synoptic development. *Quarterly Journal of the Royal Meteorological Society*, **106**, 707–19.

IPCC, 1990 *Climate Change 1990: The IPCC Scientific Assessment*. Cambridge: Cambridge University Press, 366 pp.

IPCC, 1992 *Climate Change 1992: IPCC Supplement*. Cambridge: Cambridge University Press, 200 pp.

IPCC, 1996 *Climate Change 1995: The IPCC Scientific Assessment*. Cambridge: Cambridge University Press, 572 pp.

Jones, D. A. 1994 An objective study of southern hemisphere synoptic activity. PhD Thesis, University of Melbourne, 304 pp.

Keyser, D., Reeder, M. J. and Reed, R. J. 1988 A generalization of Petterssen's frontogenesis function and its relation to the forcing of vertical motion. *Monthly Weather Review*, **116**, 762–80.

Kondatyev, K. Ya. 1969 *Radiation in the atmosphere*. New York: Academic Press.

Lindzen, R. S. 1991 Review of Climate Change, The IPCC Scientific Assessment. *Quarterly Journal of the Royal Meteorological Society*, 651–2.

Love, B. I. 1985 Cross equatorial influence of winter hemisphere subtropical cold surges. *Monthly Weather Review*, **113**, 1499–1509.

Love, G. and Murphy, K. 1987 The operational analysis of tropical cyclone wind fields in the Australian Northern Region. NT Research Papers 1984–85, Bureau of Meteorology, Northern Territory, Australia, 44–51.

McIlveen, R. 1992 *Fundamentals of weather and climate*. London: Chapman and Hall, 497 pp.

Merrill, R. T. 1993 Tropical cyclone structure. Chapter 2 in Holland, G. J. (ed.) *Global guide to tropical cyclone forecasting*. WMO/TD 560, TCP Report 31.

Miller, B. I. 1958 On the maximum intensity of hurricanes. *Journal of Meteorology*, **15**, 184–95.

Neumann, C. J. 1993 Global overview. Chapter 1 in Holland, G. J. (ed.) *Global guide to tropical cyclone forecasting*, WMO/TD 560, TCP Report 31.

Palmen, E. and Newton, C. W. 1969 *Atmospheric circulation systems*. New York: Academic Press.

Partridge, G. W. and Platt, C. M. R. 1976 *Radiative processes in meteorology and climatology*. Amsterdam: Elsevier.

Pedlosky, J. 1987 *Geophysical Fluid Dynamics*, 2nd edn. New York: Springer-Verlag, 710 pp.

Petterssen, S. 1956 *Weather forecasting and analysis*, 2nd edn. New York: McGraw-Hill.

Phillips, N. A. 1956 The general circulation of the atmosphere, a numerical experiment. *Quarterly Journal of the Royal Meteorological Society*, 123–64.

Platzman, G. W. 1990 Charney's recollections. In Lindzen, R. S., Lorenz, E. N. and Platzman, G. W. (eds) *A challenge: the science of Jule Gregory Charney*. Boston: AMS, 11–82.

Reddin, W. J. 1985 *The best of Bill Reddin*. London: Institute of Personnel Management, 180 pp.

Reeder, M. J. and Smith, R. K. 1992 Australia spring and summer cold fronts. *Australian Meteorological Magazine*, **41**, 101–24.

Riehl, H. 1978 *Introduction to the atmosphere*. Tokyo: McGraw-Hill.

Saucier, W. J. 1955 *Principles and practice of synoptic analysis*. Chicago: University of Chicago Press.

Schubert, W. H. and Hack, J. J. 1982 Inertial stability and tropical cyclone development. *Journal of Atmospheric Science*, **39**, 1687–97.

Schwerdtfeger, P. 1991 *Introduction to thermodynamics of the atmosphere*. Adelaide: Flinders Press.

Schwerdtfeger, P. 1995 Radiation in the atmosphere. Flinders University of South Australia.

Scorer, R. S. 1978 *Environmental aerodynamics*, Ellis Horwood series in mathematics and its applications. Chichester: Ellis Horwood.

Simpson, G. C. 1928 Further studies in terrestrial radiation. *Memoirs of the Royal Meteorological Society*, **3**, no. 21.

Smigielski, J. and Mogil, H. Michael 1992 A systematic approach for estimating central pressures of mid-latitude oceanic storms. NOAA Technical Report NESDIS 63.

Spencer, R. W. and Christy, J. R. 1990 Precise monitoring of global temperatures from satellites. *Science* **247**, 1558–62.

Sutcliffe, R. C. 1947 A contribution to the problem of development. *Quarterly Journal of the Royal Meteorological Society*, **73**, 370–83.

Trenberth, K. E. 1978 On the interpretation of the diagnosis quasi-geostrophic omega equation. *Monthly Weather Review*, **106**, 131–7.

UKMO 1964 *Handbook of weather forecasting*, vol. 1. United Kingdom Meteorological Office.

Wells, N. 1986 *The atmosphere and ocean: a physical introduction*. London: Taylor & Francis.

WMO 1995 *Global perspectives on tropical cyclones*. WMO/TD 693, TCP Report 38.

Young, M. V., Monk, G. A. and Browning, K. A. 1987 Interpretation of satellite imagery of a rapidly deepening cyclone. *Quarterly Journal of the Meteorological Society*, **113**, 1089–116.

Current Australian weather:

http://www.bom.gov.au/weather/national/charts/
http://www.bom.gov.au/weather/national/satellite

El Nino, SOI, sea surface temperature, Australian rainfall:

http://www.dpi.qld.gov.au/longpdk/
M.s.l. charts (Australia):

Interactive meteorology, by Richard Lowe (Curtin University of Technology, Perth, Western Australia). Distinctive characteristics and behaviour of a number of key features in a typical summer-type m.s.l. chart pattern for Australia.

Index

Page numbers in **bold** type refer to figures